NUCLEAR MATERIALS AND DISASTER RESEARCH

# NUCLEAR SAFETY IN THE WAKE OF THE FUKUSHIMA DAIICHI ACCIDENT

## ACTIONS OF SELECTED COUNTRIES

# NUCLEAR MATERIALS AND DISASTER RESEARCH

Additional books in this series can be found on Nova's website under the Series tab.

Additional e-books in this series can be found on Nova's website under the e-book tab.

NUCLEAR MATERIALS AND DISASTER RESEARCH

# NUCLEAR SAFETY IN THE WAKE OF THE FUKUSHIMA DAIICHI ACCIDENT

## ACTIONS OF SELECTED COUNTRIES

GLENN FREEMAN
EDITOR

nova publishers
*New York*

**Library of Congress Cataloging-in-Publication Data**

ISBN: 978-1-63482-280-0

*Published by Nova Science Publishers, Inc. † New York*

# CONTENTS

# PREFACE

This book examines the actions nuclear regulatory bodies from selected countries have taken to strengthen nuclear safety; the extent to which these countries have established automated systems to collect and transmit accident data; and steps international organizations have taken to support nuclear regulatory bodies and promote nuclear safety worldwide since the accident.

Chapter 1 – The March 2011 accident at Japan's Fukushima Daiichi nuclear power plant led to a worldwide review of nuclear power programs. NRC licenses and oversees civilian nuclear reactors. The State Department coordinates policy matters with international organizations and treaties, including those dealing with nuclear safety.

GAO was asked to examine (1) the actions nuclear regulatory bodies from selected countries have taken to strengthen nuclear safety; (2) the extent to which these countries have established automated systems to collect and transmit accident data; and (3) steps international organizations have taken to support nuclear regulatory bodies and promote nuclear safety worldwide since the accident. The countries GAO selected represent a cross section of established and emerging nuclear power countries. GAO also reviewed relevant documents and interviewed or obtained information from U.S. federal agencies, 15 foreign nuclear regulatory bodies, and international organizations.

Chapter 2 – The huge earthquake and tsunami that struck Japan's Fukushima Daiichi nuclear power station on March 11, 2011, knocked out backup power systems that were needed to cool the reactors at the plant, causing three of them to undergo fuel melting, hydrogen explosions, and radioactive releases. Radioactive contamination from the Fukushima plant forced the evacuation of communities up to 25 miles away and affected up to 100,000 residents, although it did not cause any immediate deaths.

Tokyo Electric Power Company (TEPCO) operates the Fukushima nuclear power complex in the Futaba district of Fukushima prefecture in Northern Japan, consisting of six nuclear units at the Fukushima Daiichi station and four nuclear units at the Fukushima Daini station. All the units at the Fukushima complex are boiling water reactors, with reactors 1 to 5 at the Fukushima Daiichi site being the General Electric Mark I design, which is also used in the United States. The Fukushima Daiichi reactors entered commercial operation in the years from 1971 (reactor 1) to 1979 (reactor 6). The Fukushima Daini reactors shut down automatically after the earthquake and were able to maintain sufficient cooling.

When the earthquake struck, Fukushima Daiichi units 1, 2, and 3 were generating electricity and shut down automatically. The earthquake caused offsite power supplies to be lost, and backup diesel generators started up as designed to supply backup power. However, the subsequent tsunami flooded the electrical switchgear for the diesel generators, causing most AC power in units 1 to 4 to be lost. Because Unit 4 was undergoing a maintenance shutdown, all of its nuclear fuel had been removed and placed in the unit's spent fuel storage pool. One generator continued operating to cool units 5 and 6.

The loss of all AC power in units 1 to 3 prevented valves and pumps from operating that were needed to remove heat and pressure that was being generated by the radioactive decay of the nuclear fuel in the reactor cores. As the fuel rods in the reactor cores overheated, they reacted with steam to produce large amounts of hydrogen, which escaped into the unit 1, 3, and 4 reactor buildings and exploded (the hydrogen that exploded in Unit 4 is believed to have come from Unit 3). The explosions interfered with efforts by plant workers to restore cooling and helped spread radioactivity. Cooling was also lost in the reactors' spent fuel pools, although recent analysis has found that no significant overheating took place.

Radioactive material released into the atmosphere produced extremely high radiation dose rates near the plant and left large areas of land uninhabitable, especially to the northwest of the plant. Contaminated water from the plant was discharged into the sea, creating international controversy.

The United States and other countries, as well as the International Atomic Energy Agency, are providing assistance to Japan to deal with the nuclear disater. U.S. assistance has included transport of pumps, boron, fresh water, remote cameras, use of Global Hawk surveillance drones, evacuation support, medical support, and decontamination and radiation monitoring equipment.

Studies of the Fukushima disaster have identified design changes, response actions, and other safety improvements that could have reduced or eliminated the amount of radioactivity released from the plant. As a result, Fukushima has prompted a reexamination of nuclear plant safety requirements around the world, including in the United States.

Chapter 3 – Nuclear energy issues facing Congress include reactor safety and regulation, radioactive waste management, research and development priorities, federal incentives for new commercial reactors, nuclear weapons proliferation, and security against terrorist attacks.

The earthquake and resulting tsunami that severely damaged Japan's Fukushima Daiichi nuclear power plant on March 11, 2011, raised questions in Congress about the disaster's possible implications for nuclear safety regulation, U.S. nuclear energy expansion, and radioactive waste policy. The tsunami knocked out electric power at the six-reactor plant, resulting in the overheating of several reactor cores, loss of cooling in spent fuel storage pools, major hydrogen explosions, and releases of radioactive material into the environment. The Nuclear Regulatory Commission (NRC) issued orders to U.S. nuclear plants March 12, 2012, to begin implementing safety improvements in response to Fukushima.

Significant incentives for new commercial reactors were included in the Energy Policy Act of 2005 (EPACT05, P.L. 109-58), such as tax credits and loan guarantees. Together with volatile fossil fuel prices and the possibility of greenhouse gas controls, the federal incentives for nuclear power helped spur renewed interest by utilities and other potential reactor developers. License applications for as many as 31 new reactors were announced, and NRC issued licenses for four reactors at two plant sites in early 2012. However, falling natural gas prices, safety concerns raised by the Fukushima accident, and other changing circumstances have made it unlikely that many more of the proposed nuclear projects will move toward construction in the near term.

Four U.S. reactors were permanently closed in 2013, and another shutdown has been announced for late 2014. Three reactors were closed because of the need for major repairs, and the other two because electricity prices fell below their generating costs.

DOE's nuclear energy research and development program includes advanced reactors, fuel cycle technology and facilities, and infrastructure support. The Obama Administration's FY2015 funding request for nuclear energy research and development totaled $863.4 million. Including advanced reactors, fuel cycle technology, infrastructure support, and safeguards and security, the total nuclear energy request was $25.0 million (2.9%) below the

FY2014 funding level. The House approved $826.7 million (H.R. 4923), while the Senate Subcommittee on Energy and Water Development Appropriations recommended $777.0 million.

Disposal of highly radioactive waste has been one of the most controversial aspects of nuclear power. The Obama Administration halted work on a long-planned waste repository at Yucca Mountain, NV, and established the Blue Ribbon Commission on America's Nuclear Future (BRC) to recommend new approaches to the waste problem. The BRC issued its final report to the Secretary of Energy on January 26, 2012. In response to the BRC report, and to provide an outline for a new nuclear waste program, DOE issued a *Strategy for the Management and Disposal of Used Nuclear Fuel and High-Level Waste* in January 2013. The DOE strategy calls for a new nuclear waste management entity to develop consent-based storage and disposal sites, similar to recommendations by the BRC. No funding has been requested or appropriated since FY2011 to continue NRC licensing of the Yucca Mountain repository, although the House voted to provide DOE with $150.0 million and NRC $55.0 million for Yucca Mountain licensing activities in FY2015.

In: Nuclear Safety ...
Editor: Glenn Freeman

ISBN: 978-1-63482-280-0


*Chapter 1*

# NUCLEAR SAFETY: COUNTRIES' REGULATORY BODIES HAVE MADE CHANGES IN RESPONSE TO THE FUKUSHIMA DAIICHI ACCIDENT*

***United States Government Accountability Office***

## WHY GAO DID THIS STUDY

The March 2011 accident at Japan's Fukushima Daiichi nuclear power plant led to a worldwide review of nuclear power programs. NRC licenses and oversees civilian nuclear reactors. The State Department coordinates policy matters with international organizations and treaties, including those dealing with nuclear safety.

GAO was asked to examine (1) the actions nuclear regulatory bodies from selected countries have taken to strengthen nuclear safety; (2) the extent to which these countries have established automated systems to collect and transmit accident data; and (3) steps international organizations have taken to support nuclear regulatory bodies and promote nuclear safety worldwide since the accident. The countries GAO selected represent a cross section of established and emerging nuclear power countries. GAO also reviewed

* This is an edited, reformatted and augmented version of a United States Government Accountability Office publication, No. GAO-14-109, dated March 2014.

relevant documents and interviewed or obtained information from U.S. federal agencies, 15 foreign nuclear regulatory bodies, and international organizations.

## WHAT GAO RECOMMENDS

GAO recommends (1) that State and NRC work with and encourage IAEA to systematically track the status of recommendations made by IAEA peer review missions and (2) NRC consider expediting its decision on whether or how to upgrade its automated system for transmitting key reactor data. NRC neither agreed nor disagreed with the recommendations. State partially concurred with the first recommendation and had no comment on the second. GAO believes that fully implementing these recommendations would enhance nuclear safety.

## WHAT GAO FOUND

All the nuclear regulatory bodies in the 16 selected countries in GAO's review— 13 of which currently operate nuclear power reactors and 3 of which are developing or considering developing civilian nuclear power programs—have taken steps to strengthen nuclear safety in response to the Fukushima Daiichi accident in Japan. Japan in particular has fundamentally restructured its nuclear regulatory framework, and 3 other countries—China, Sweden, and Vietnam—are providing additional resources to their nuclear regulatory bodies. Countries are taking steps to improve safety with a focus on considering previously unimagined accident scenarios. Specifically, regulatory bodies in several countries (e.g., Belgium, Canada, Russia, and the United States) are now planning for accident scenarios that could involve multiple reactors at a single power plant. In addition, new requirements for emergency equipment, such as backup electric generators, in case of the loss of off-site power, as occurred at the Fukushima Daiichi nuclear power plant, are an area of focus among the regulatory bodies in GAO's review.

Officials from 6 of the 13 countries with operating nuclear power reactors in GAO's review said they have automated systems for collecting and transmitting critical nuclear power plant data to the nuclear regulatory body or designated technical experts who work with the regulatory body during an accident, and officials from a seventh country said that it has plans to build such a system. Officials from 3 of the countries with automated systems, including the United States, told GAO they are considering steps to ensure their systems can operate in certain emergency conditions, such as during the loss of off-site power, but none has a specific timetable for doing so. For example, the U.S. Nuclear Regulatory Commission (NRC) is first completing higher priority nuclear safety enhancements before deciding whether or how to upgrade its automated system because how enhancements are done may affect how upgrades to an automated system would be implemented. By delaying its decision on upgrades to enable the system to function under emergency conditions, the system may not function when needed most—during a severe accident.

Three key international organizations—the International Atomic Energy Agency (IAEA), the World Association of Nuclear Operators, and the European Union— along with the Convention on Nuclear Safety, have taken steps to support nuclear regulatory bodies and help them identify the most important lessons of the Fukushima Daiichi accident and promote regulatory changes to enhance nuclear safety worldwide. For example, one key way IAEA helps countries improve nuclear safety and regulatory effectiveness is through peer review missions, which evaluate, among other things, a country's nuclear safety regulatory framework based on IAEA Safety Standards and good regulatory practices. However, according to IAEA officials, the agency does not systematically track whether the recommendations of the peer review missions are implemented by the host countries. Without this information, IAEA cannot fully determine the impact and effectiveness of the peer review missions.

## ABBREVIATIONS

| | |
|---|---|
| Action Plan | Action Plan on Nuclear Safety |
| ASN | Autorité de Sûreté Nucléaire |
| CNSC | Canadian Nuclear Safety Commission |
| Convention | Convention on Nuclear Safety |
| DOE | Department of Energy |
| ENS | Emergency Notification System |
| ENSREG | European Nuclear Safety Regulators Group |
| ERDS | Emergency Response Data System |
| EU | European Union |

| | |
|---|---|
| EURATOM | European Atomic Energy Community |
| IAEA | International Atomic Energy Agency |
| IRRS | Integrated Regulatory Review Service |
| IRRT | International Regulatory Review Team |
| NEA | Nuclear Energy Agency |
| NEPC | Nuclear Emergency Preparedness Commission |
| NISA | Nuclear and Industrial Safety Agency |
| NRA | Nuclear Regulation Authority |
| NRC | U.S. Nuclear Regulatory Commission |
| NSSC | Nuclear Safety and Security Commission |
| OECD | Organization for Economic Cooperation and Development |
| ONR | Office for Nuclear Regulation |
| OSART | Operational Safety Review Team |
| SSM | Swedish Radiation Safety Authority |
| TEPCO | Tokyo Electric Power Company |
| UN | United Nations |
| VPN | Virtual Private Network |
| WANO | World Association of Nuclear Operators |
| WENRA | West European Nuclear Regulators' Association |

* * *

March 6, 2014

The Honorable Thomas R. Carper
Chairman
Subcommittee on Transportation and Infrastructure
Committee on Environment and Public Works
United States Senate

Dear Mr. Chairman:

The March 2011 accident at Japan's Fukushima Daiichi nuclear power plant destroyed three nuclear reactors and resulted in the most extensive release of radioactive material at a nuclear power plant since the 1986 Chernobyl disaster.[1] The accident, which led to the evacuation of over one hundred thousand residents from the area around the plant and is expected to cost Japan tens of billions of dollars, has led to a review of civilian nuclear power programs worldwide. For example, Germany accelerated the shutdown of its nuclear power reactors, and Jordan reassessed plans to establish a civilian nuclear power program.[2] Nuclear regulators and industry officials question whether public acceptance of civilian nuclear power could survive another severe accident. The Secretary General of the United Nations (UN) noted at a September 2011 UN high-level summit on nuclear safety and security that "the effects of nuclear accidents respect no borders."

Any event that compromises a nuclear power plant's power supplies can create the conditions for a nuclear accident. The Fukushima Daiichi accident resulted from a prolonged loss of electrical power when a powerful earthquake triggered a tsunami wave that exceeded the plant's seawall and flooded the site. The loss of power also largely disabled Japan's automated system for collecting and transmitting data during emergencies. Various stakeholders, including Tokyo Electric Power Company (TEPCO), which operated the Fukushima Daiichi power plant, acknowledged that design and operational shortcomings contributed to the accident sequence. Furthermore, numerous studies have concluded that failures in regulation contributed to those shortcomings. For example, according to the Kurokawa Commission[3]—an investigation commissioned by Japan's parliament—Japan's nuclear regulatory body at the time of the accident, the Nuclear and Industrial Safety Agency (NISA), and TEPCO were aware of the risks of a station blackout and reactor core damage from the loss of seawater pumps from a tsunami. According to the investigation, NISA knew that TEPCO had not taken actions to mitigate those risks but did not require any corrective measures. The Kurokawa Commission attributed NISA's reluctance to require greater preparedness of TEPCO to a flawed safety culture and to the fact that NISA was part of the government ministry responsible for promoting nuclear power, finding that NISA was not independent of the nuclear industry that it was charged with regulating or of the government agency responsible for promoting that industry.[4] NISA's lack of independence is believed by the Kurokawa Commission to have contributed to NISA's lax oversight of the nuclear industry. Nuclear industry representatives have said that their industry needs effective nuclear safety regulation and that the regulatory body must be independent from both operators and politics. The international cooperative and assistance programs of the U.S. Nuclear Regulatory Commission (NRC)—the nuclear regulatory body of the United States—over the past 20 years have focused on promoting the independence and effectiveness of such bodies.

Nuclear power reactors generated about 11.3 percent of the world's electricity in 2012. As of November 2013, there are 435 civilian nuclear power reactors operating in 30 countries and 71 more under construction, primarily in China and Russia. In addition, as we reported in November 2010, countries such as Vietnam and Jordan, which do not yet have operating civilian nuclear power reactors, are building the necessary regulatory infrastructure for nuclear programs.[5] In April 2013, international regulators from 49 countries gathered in Ottawa, Canada, for a conference organized by the International Atomic Energy Agency (IAEA) devoted to improving nuclear regulatory systems following the Fukushima Daiichi accident.[6] The conference concluded that nuclear regulatory bodies have learned many lessons from the Fukushima Daiichi accident and that implementing these lessons will take time and commitment.

In light of the Fukushima Daiichi accident, you asked us to examine worldwide nuclear regulation and safety.[7] This report examines: (1) the actions regulatory bodies from selected countries with existing or planned civilian nuclear power reactors have taken to strengthen nuclear safety; (2) the extent to which these countries have established automated systems for collecting and transmitting data to the nuclear regulatory body and taken steps to enable such systems to withstand emergency conditions; and (3) steps key international organizations have taken to support nuclear regulatory bodies and promote nuclear safety worldwide since the Fukushima Daiichi accident and the impact of some of these steps.

To examine the actions that national regulatory bodies have taken to strengthen nuclear safety in the aftermath of the Fukushima Daiichi accident, we selected a nonprobability sample of 16 countries for case studies to represent a cross section of established and emerging civilian nuclear power countries in terms of program size,[8] the country's dependence on nuclear power, and the country's nuclear regulatory framework.[9] Of these 16 countries, the 13 countries with established nuclear power programs operate 78 percent of the world's nuclear power reactors; the other 3 countries do not currently operate nuclear power reactors but have emerging civilian nuclear power programs or have stated an interest in developing such programs.[10] Appendix II contains a complete list of these countries and information on their civilian nuclear power programs. For the 16 countries selected, we reviewed reports they prepared to document the lessons learned from the Fukushima Daiichi accident, including reports prepared for various international meetings on nuclear safety, and/or presentations in which they summarized those lessons.[11] We interviewed one or more officials from each selected regulatory body and/or technical support organization and/or received written responses to our questions. For some countries, we interviewed nuclear power plant operators and/or other nuclear energy industry officials. We also interviewed NRC and U.S. nuclear industry officials, as well as two former NRC chairmen. To examine the extent to which other countries have established or are considering establishing automated systems for collecting and transmitting data to the nuclear regulatory body during emergencies, we interviewed officials or received written answers to our questions from the respective nuclear regulatory bodies and some power plant operators. Because some country officials were more responsive than others, the amount of corroborating evidence we were able to obtain varies by country. For more information, see appendix I. To examine the steps key international organizations have taken to support nuclear regulatory bodies and promote nuclear safety worldwide since the Fukushima Daiichi accident, we first identified such organizations through interviews with officials from the U.S. State Department, Department of Energy (DOE), and NRC officials. Based on these discussions, we identified IAEA, the Organization for Economic Cooperation and Development's (OECD)[12] Nuclear Energy Agency (NEA),[13] the World Association of Nuclear Operators (WANO),[14] and the European Union (EU). We then reviewed, as applicable, the reports or action plans of these organizations and interviewed or exchanged written questions with relevant officials from them.

We conducted this performance audit from May 2012 to March 2014 in accordance with generally accepted government auditing standards. Those standards require that we plan and perform the audit to obtain sufficient, appropriate evidence to provide a reasonable basis for our findings and conclusions based on our audit objectives. We believe that the evidence obtained provides a reasonable basis for our findings and conclusions based on our audit objectives. Appendix I contains more detailed information on our scope and methodology.

## BACKGROUND

This section describes (1) the Fukushima Daiichi nuclear accident, (2) nuclear energy and its regulation, (3) automated data transmission systems used by some nuclear regulatory bodies, and (4) international organizations and a treaty involved in promoting nuclear safety.

## The Fukushima Daiichi Nuclear Accident

The March 11, 2011, tsunami wave that flooded the Fukushima Daiichi nuclear station destroyed the station's power supplies and emergency equipment, which, in turn prevented the plant's operators from cooling the reactors. The tsunami wave, at 13.1 meters, or nearly 43 feet was beyond the Fukushima Daiichi plant's "design basis"—that is, it exceeded the size of the tsunami that the site was designed and built to withstand, which was 3.1 meters, or slightly more than 10 feet above sea level. The station's emergency power supplies, including diesel generators and batteries, were below ground and not in waterproof containers. The station's three operating reactors had shut down during the earthquake that preceded the tsunami,[15] but the radioactive fuel continued to decay and produce heat. The tsunami destroyed the Fukushima Daiichi power station's emergency power supplies, with the exception of one diesel generator for the sixth reactor.[16] The electric wiring system for the three operating reactors, as well as seawater pumps, which were used to remove heat from the plant and cool diesel generators, were also destroyed. Without power to cool the reactors, the water in the reactor vessels boiled away and exposed the fuel, which melted. The steam from the boiling water increased pressure inside the primary containment vessel, allowing steam and other gases to move into the reactor building.[17] The steam reacted with melting fuel, generating hydrogen gas that built up in at least two of the reactor buildings before eventually exploding. In the wake of the accident, the Japanese government has directed that all but 2 of Japan's 50 civilian nuclear power reactors be shut down pending a complete safety review and, as of this writing, all reactors have been shut down. It is uncertain when these reactors will be brought back online. Figure 1 shows Unit 3 of the Fukushima Daiichi nuclear power station following the March 11, 2011, earthquake and tsunami.

## Nuclear Energy and Its Regulation

The safe operation of nuclear power reactors worldwide has been a long-standing concern of the international community. According to IAEA, it is a fundamental safety principle that nuclear power plant operators have primary responsibility for nuclear safety during the life of the plant. However, nuclear regulatory bodies play a key role in ensuring nuclear safety. In the United States, NRC, an independent federal agency composed of five commissioners, licenses civilian nuclear power reactors and regulates and oversees their safe operation and security.

According to IAEA documents, effective nuclear regulatory bodies are characterized by independence, as well as other key factors, such as transparency and commitment to safety culture. NRC, for example, defines regulatory independence as independence from

- economic interests regarding the use of any nuclear materials or technology,
- policy interests, and
- political interests.

To that end, according to NRC, it does not (1) make decisions based on financial costs to nuclear plant operators when considering matters of safety significance or (2) consider the impact of its actions on the future of nuclear energy or any particular technology. Furthermore, NRC does not report to a cabinet agency or to the White House.

Source: NRC

Figure 1. Unit 3 of the Fukushima Daiichi Nuclear Power Plant Following the March 11, 2011, Earthquake and Tsunami.

## Automated Data Transmission Systems

Automated systems for collecting and transmitting data enable operators to provide data on selected nuclear power plant parameters—such as the status of reactor coolant and containment systems, radiation monitoring and containment, and weather—directly to regulators. According to NRC officials, receiving automated, accurate, real-time data directly from automated sources at the site reduces the burden placed on operators to manually transmit information during an emergency and gives confidence to government authorities involved in responding to the event and the public that the plant operator is not withholding information.

## International Organizations and Treaty Involved in Promoting Nuclear Safety

IAEA, WANO, and the EU play an important role in promoting nuclear safety worldwide. Specifically:

- IAEA develops (nonbinding) international standards for nuclear safety, as well as criteria for nuclear regulatory independence. IAEA also offers technical cooperation and peer review missions to its member states. These peer review missions evaluate, among other things, a country's nuclear safety regulatory framework based on the IAEA Safety Standards and conduct in-depth reviews of operational safety performance at nuclear power plants. A Secretariat, headed by the Director General, is responsible for implementing the agency's policies and programs. The U.S. State Department is the lead agency on U.S. policy regarding the IAEA, including the promotion of effective functioning in the agency and management reform. In addition, the department coordinates U.S. policy matters with international organizations and treaties, including those dealing with nuclear safety. NRC participates regularly in a variety of IAEA activities related to, among other things, nuclear safety. For example, NRC serves as the U.S. government's representative to the IAEA Commission on Safety Standards, as well as other technical committees.
- WANO, established in 1989, 3 years after the Chernobyl accident, promotes the safe and reliable operations of all of the world's civilian

nuclear power plants. WANO works to achieve its goals through confidential peer reviews, shared access to members' operating experience, technical support and exchange activities, and professional and technical development sessions. These peer reviews help members compare their operational performance against standards of excellence through an in-depth, objective, and confidential review of their operations conducted by nuclear experts drawn from other WANO members' plants. To permit the free and open exchange of information among WANO members, plant specific information, such as that gathered in the course of a peer review, may not be released outside WANO without the approval of the originating member. Since 1992, WANO has conducted more than 500 peer reviews of operating nuclear power plants in 31 countries or regions— including at least one at every WANO member plant. Membership is voluntary, but every nuclear operator in the world is currently a member, and participation in these activities is mandatory for members. WANO members are committed to take timely action to correct performance issues identified during peer reviews.

- The EU was active in the aftermath of the Fukushima accident, exploring and implementing options for using its authorities to advance nuclear safety in Europe, particularly among the 14 EU member states with operating nuclear power plants. The requirements in the EU nuclear safety directive and nuclear waste directive are legally binding in all EU member states.

The Convention on Nuclear Safety (the Convention) is a multilateral treaty that, among other things, seeks to strengthen the safety of civilian nuclear power reactors. In the mid-1990s, representatives of over 50 nations, including the United States, participated in the development of the Convention in the aftermath of the Chernobyl accident. Today, 75 countries, including the United States, and one international organization, the European Atomic Energy Community (EURATOM), are contracting parties to the Convention.[18] The Convention calls on parties to establish and maintain a legislative and regulatory framework and a regulatory body with adequate authority, competence, and financial and human resources to govern the safety of nuclear installations, among other things. Specifically, parties to the Convention are obligated to take steps to ensure an effective separation between the regulatory body and any other body or organization concerned with the promotion or utilization of nuclear energy. Under the terms of the Convention, each contracting party— regardless of whether it operates nuclear power plants—is obligated to submit a national report, months in advance of the review meeting, which identifies the measures taken to implement each of the relevant nuclear safety obligations contained in the Convention. Obligations cover such points as siting, design, construction, and operation of civilian nuclear power installations. Parties that do not operate nuclear power plants submit reports focusing on, among other things, the steps they have taken to prepare and test emergency plans to deal with an accident in a neighboring country that operates nuclear power plants. Countries considering establishing nuclear power programs are expected to provide information in their national reports about the steps they are taking to meet the Convention's obligations including, for example, reactor design and siting requirements. The parties to the Convention have also established detailed guidance to help parties prepare their national reports. The purpose of the guidance is to encourage parties to describe the steps they are taking to meet the Convention's obligations and to facilitate other parties' review of the national reports of other countries. The countries meet every 3 years in Vienna, Austria, for review meetings to present their national report, address questions that may arise about the report, and assess and ask questions about the reports of other parties. This is considered a peer review process. Parties may also submit written questions in advance of the review meeting on other parties' national reports and receive written responses to these questions.

## Nuclear Regulatory Bodies Have Taken Steps to Strengthen Safety

The nuclear regulatory bodies in the selected 16 countries in our review have taken steps to strengthen nuclear safety. Countries are taking steps with an increased focus on considering previously unimagined accident scenarios, such as those affecting multiple reactors within a given power plant. In addition, most countries with operating reactors in our review are considering or have issued new requirements in common categories: emergency equipment, hydrogen control, and filtered venting. Japan has fundamentally restructured its regulatory framework, and 3 other countries—China, Sweden, and Vietnam—are providing their nuclear regulatory bodies with additional resources. Four countries—France, South Korea, Sweden, and the United

Kingdom—have also reformed their regulatory framework but not specifically in response to the Fukushima Daiichi accident.

## Countries Are Strengthening Regulatory Effectiveness and Considering Previously Unimagined Events and Accident Scenarios

All 16 selected countries in our review have taken regulatory action in response to the Fukushima Daiichi accident. For example, in the United States, NRC formed the Near-Term Task Force in March 2011 to conduct a comprehensive assessment of NRC's processes and regulations in light of the accident to determine whether the agency should make additional improvements to its regulatory system or reactor safety in the United States. This task force issued a report in July 2011 with its findings and 12 recommendations, which were prioritized in an October 2011 report into three tiers. The commission then approved the NRC staff's three-tier prioritization in December 2011. Tier 1 recommendations were those that NRC staff determined should be started without unnecessary delay and for which resources are sufficient. Tier 2 comprises recommendations requiring further technical assessment or that depended on Tier 1 issues or availability of resources. Tier 3 recommendations require further study, depend on a shorter-term action to be completed to inform the longer-term action, or depend on the availability of resources or on the resolution of Task Force Recommendation 1, which deals with clarification of the regulatory framework. All of the items identified for long-term evaluation fall into Tier 3.

Each of the other selected countries in our review, with the exception of Vietnam, which does not have operating power reactors—undertook safety inspections, including comprehensive risk safety assessments, more commonly known as stress tests.[19] The findings from the stress tests were used to consider further action to improve nuclear safety. Stress tests examine what are known as "beyond design-basis events"— those where conditions exceed what the facility or site was designed to withstand. For example, one of the beyond design-basis scenarios considered by the Belgian Doel nuclear power plant operator was a severe storm combined with an unfavorable wind direction—which together could breach the embankment protecting the plant from a flood.[20] For this scenario, the plant's operator analyzed the consequences of such a breach, estimating flooding levels and potential consequences, including weak points—such as buildings not designed to be watertight— as well as ways in which the plant was adequately prepared, such as emergency equipment that remains protected.

The results of the stress tests served as a basis for corrective measures. For example, the proposed corrective measure for the scenario described at the Doel plant is to reinforce the top of the embankment with concrete tiles. In Sweden, regulators are considering a revision of the country's emergency planning zones after stress tests revealed that the existing emergency planning zones were insufficient for a radioactive release similar to that of the Fukushima Daiichi accident. 21 In the United Arab Emirates, which does not have operating power reactors, the regulatory body commissioned a task force to study the implications of the Fukushima Daiichi accident from the standpoint of facilities with pending license applications, where operators were asked to supplement their license applications in consideration of external event contingencies, according to regulatory officials we interviewed. In the United States, NRC did not require plant operators to conduct stress tests. Instead, nuclear power plant operators were required to complete detailed inspections and engineering assessments of their plants, known as "walk-downs," to check that all safety features were in operating condition and to submit reports to NRC for review. Any issues identified in the walk-downs were to be addressed by plants' corrective action programs.[22] For example, the flooding walk-down at the Vermont Yankee power plant identified a discrepancy in the procedure for sandbag barricades. The procedure has since been enhanced to provide specific, detailed guidance on the placement of sandbag barriers, and an exercise in accordance with the new procedure was successfully completed.

According to a 2013 IAEA nuclear safety review, nuclear regulatory bodies are now focusing more on a wider range of possible events as a result of the Fukushima Daiichi accident. According to this review, nuclear regulatory bodies have not historically required nuclear operators to include multireactor accidents in their design basis. Canadian Nuclear Safety Commission (CNSC) officials also told us that a significant shift in regulatory focus in Canada pertains to the management of multiple reactors. They said that prior to the Fukushima Daiichi accident, which destroyed four reactors, CNSC's guidelines did not consider the possibility of an event that could impact multiple reactors. These officials told us that certain safety features are more complicated for multireactor plants because some systems and equipment are shared among reactors, allowing for common-cause failures during an accident. CNSC officials told us that, in response to the Fukushima Daiichi accident, they are incorporating multireactor specifications into their regulations. Similarly, a Belgian nuclear regulatory official said that the Fukushima Daiichi accident has led regulatory bodies to "think outside the box" in considering scenarios that could threaten nuclear power plants. For example, one of the items in a national action plan that the Belgian nuclear regulatory body developed after the accident pertains to an internal emergency plan in case of an event that impacts multiple reactors. A Russian regulatory official also told us that regulators need to consider previously unimagined events, including the impact on nuclear power plants from a meteor, such as the one that struck Russia in February 2013. [23] In March 2012, in keeping with the shift toward considering events that may impact multiple reactors, Russia's nuclear regulatory body approved a corrective action plan that includes special consideration for plants with multiple reactors.[24] In the United States, NRC is revisiting its beyond design-basis requirements to better integrate them into its overall regulatory procedures. For example, NRC's Near-Term Task Force report contains a section that focuses on operators' capabilities to respond to a prolonged station blackout—that is, a loss of power—and events that impact multiple reactors. This report explains that the accident at Fukushima Daiichi has shown that such events are realities that must be addressed as part of emergency planning because, although they are of low probability, they have the potential for severe consequences.[25] In that vein, an NRC commissioner said, in June 2013, that the accident has blurred the distinction between design-basis and beyond design-basis.

## Most Countries in Our Review Are Considering Common Categories of New Regulatory Requirements — Emergency Equipment, Hydrogen Control, and Filtered Venting Systems

To improve nuclear safety based on the lessons learned from the Fukushima Daiichi accident, the regulatory bodies in our review are considering or have issued new requirements with regard to the following common categories: emergency equipment, hydrogen control, and filtered venting systems. In the course of the Fukushima Daiichi accident, the loss of emergency equipment—including backup power supplies, such as batteries and diesel generators, as well as water supplies, such as pumps—prevented operators from cooling the reactors. Consequently, as the water in the reactors boiled and the fuel melted, hydrogen built up and exploded, destroying three reactor containment structures and releasing radioactivity into the environment. New regulatory requirements worldwide in response to these events include the following:

- *Emergency equipment.* New requirements for emergency equipment—for example, the number of diesel generators on-site and their placement—are an area of focus among the regulatory bodies in our sample. Examples are as follows:
- South Korea's regulatory body issued a list of 50 action items, which included securing the availability of a portable electric power generator vehicle and batteries and installing paths for the injection of emergency cooling water.
- The Canadian Nuclear Safety Commission has issued new requirements for emergency equipment and resources, such as requirements for backup power at emergency facilities; on-site storage of emergency equipment, such as portable generators; and the availability of portable backup power and emergency response equipment.
- NRC has issued an order to address operators' capabilities to move water to maintain or restore reactor core cooling, among other things. A senior NRC official said that, because the order requires these capabilities to be maintained even if power is lost, the order includes implicit requirements for emergency power supplies or suitable alternative methods to continue reactor core cooling indefinitely.
- *Hydrogen control.* All but one of the 13 nuclear regulatory bodies in our review with operating nuclear power reactors in their countries have issued additional requirements for hydrogen control systems or are studying alternative options for hydrogen removal. In some cases, countries that already have some hydrogen control systems are looking to test existing systems and further strengthen hydrogen safety. For example, Russia's regulatory body required complete implementation of hydrogen monitoring and removal systems in reactors where they were not part of the original design and also required an analysis of the adequacy of existing hydrogen monitoring and removal systems to prevent the formation of explosive gas mixtures in the most unfavorable scenarios in severe accidents. Similarly, France's regulatory body has required that one type of hydrogen control system be qualified for external hazards beyond the level currently considered,[26] and has issued requirements for the detection of hydrogen in places not planned for in the design.

- *Filtered venting systems.* Eleven of the 13 regulatory bodies in our review with operating reactors have issued new requirements or are studying options for filtered venting systems to minimize the release of radioactive materials into the environment in the event of an accident.

For example, filtered venting systems are included in Japan's new safety regulations under "measures for preventing core damage/containment failure." Canada's Action Plan in response to the Fukushima Daiichi accident includes requirements for "Emergency Filtered Containment Venting." Countries that already have filtered venting systems, such as Sweden, are studying potential updates and improvements to their systems—for example, to ensure that they function in prolonged accident scenarios. France's regulatory body has required the operator to submit a detailed study of possible improvements to its venting-filtration system, taking into account the possibility of needing filtration in two or more reactors simultaneously. In the United States, NRC announced that it is in the process of a technical evaluation to support rulemaking on filtering, with the deadline for the final rule set for 2017. NRC staff recommended that the commission order the installation of filtered venting systems at boiling water reactors with Mark I and II containments—those similar to the ones at Fukushima Daiichi.[27] However, the commission voted against issuing an immediate order to require installation of filtered venting systems and instead gave NRC staff a year to produce a technical evaluation to support rulemaking on filtering,[28] thus deciding to address filtered venting through its regular process rather than as an emergency issue. The commission directed the staff to consider both the use of a filter to be placed on the vent, as well as a more performance-based approach using existing systems to achieve a similar reduction in radioactive release during an accident.

## Japan Restructured Its Nuclear Regulatory Framework, and China, Sweden, and Vietnam Devoted Additional Resources to Nuclear Safety

Japan was the only country in our review that fundamentally restructured its nuclear regulatory framework in response to the Fukushima Daiichi accident. Specifically, in June 2012, Japan's parliament passed the Establishment of the Nuclear Regulation Authority Act, which, effective in

September 2012, established the new Nuclear Regulation Authority (NRA) as an independent commission in charge of all of the nation's nuclear regulatory functions. NRA replaced the organizations previously responsible for nuclear safety regulation, including NISA. This restructuring separated the government's nuclear regulatory function from that of nuclear power promotion. The reform also established a Nuclear Emergency Preparedness Commission (NEPC) under Japan's cabinet and required the new regulatory body to incorporate the latest scientific and technical knowledge into the nation's nuclear regulatory basis, in keeping with the Kurokawa Commission finding that NISA had a negative attitude toward new advances in knowledge and technology from outside Japan. An NRA commissioner told us that Japan's restructuring of its nuclear regulatory system is necessary to address the issue of "regulatory capture"—the collusion between NISA and the nuclear industry—that compromised the nation's nuclear safety prior to the accident and to regain the public trust, which the commissioner told us was NRA's biggest challenge. NRA's mission statement lists among the agency's core values and principles the fostering of a genuine safety culture, where public safety is the highest priority.

According to the commissioner, NRA's structure is largely modeled on NRC's, and, to a lesser extent, on that of France's nuclear regulatory body. For example, NRA will have a five-commissioner voting system like the U.S. and French nuclear regulatory bodies. However, the Japanese government has not determined how to reconcile NRA's independence from the Japanese bureaucracy with its status as a government agency within that bureaucracy. According to the commissioner, the fragmentation under the previous system—with multiple agencies responsible for various aspects of nuclear safety— was problematic, and the need for integration "obvious" and that integrating all nuclear oversight agencies into one is an international trend. He said, however, that merging the staff and budgets of previously separate agencies will take time, as NRA is an amalgam of these disparate agencies. The commissioner said that NRA currently has no budget autonomy, as its budget, as that of any other Japanese government agency, is allocated by the government. However, the issue of NRA funding has not been finally settled, and fees from nuclear operators are still under consideration as a funding source. The commissioner said that funding the agency independently of the government will be difficult but that not having budgetary independence could limit organizational and personnel independence, which are essential.

The NRA completed new regulations for civilian nuclear power plants in Japan in July 2013. All of Japan's nuclear power plants with operators that wish to restart their reactors must meet criteria established by the new regulations. In addition, reactors built in the 1970s and 1980s will be reexamined in light of the latest science on earthquakes and geology. Once the safety of each reactor is assessed based on the new regulations—a process that is expected to take 2 to 3 years—the Japanese government is to determine whether each reactor can restart. The operator of each reactor must also obtain the support of the public; the governor of the prefecture where the reactor is located and, in some cases, the governors of nearby prefectures; and the national government of Japan. NRA will only be involved in the technical deliberations and will not have a role in other aspects of the governmental decisions to restart reactors.

Three other countries in our review—China, Vietnam, and Sweden—did not change their regulatory structures in response to the Fukushima Daiichi accident, but senior nuclear regulatory officials we interviewed from those countries said that the accident reinforced to their governments the importance of a strong regulatory body. As a result, the officials said that their governments provided additional human and financial resources to their nuclear regulatory bodies. According to the respective regulatory officials, the governments of China and Vietnam paused or slowed their nuclear programs after the Fukushima Daiichi accident to strengthen their regulatory authorities. The Chinese government suspended the licensing of new reactors in 2011—there were 14 operating reactors and additional ones under construction at that time—until a National Nuclear Safety Plan was developed and safety inspections were completed, according to a senior regulatory official. The Chinese government also agreed to nearly double the regulatory body's staff to 1,100 and increased its budget. In Vietnam, the government plans to begin construction on its first nuclear power plant in 2014 and to begin operation in 2020. However, since the Fukushima Daiichi accident, the nuclear regulatory body sees this timeline as overly ambitious as the country needs time to develop a nuclear safety infrastructure that meets IAEA standards. To that end, Vietnam's Atomic Energy Law will be reevaluated in the aftermath of Fukushima with regard to regulatory independence, among other things. According to the Department of Energy, Vietnam announced in January 2014 that the start of construction on its first nuclear power plant may be delayed until the 2017-2018 timeframe. In Sweden, senior regulatory officials told us that the government had previously hesitated to fund the regulatory body with the amount it had requested but, after the accident, it provided a larger budget allocation than requested, allowing the regulatory body to hire more staff.[29]

## Four Countries' Regulatory Changes Were Not Made in Response to the Fukushima Daiichi Accident

Four countries in our sample—France, Sweden, South Korea, and the United Kingdom—have recently made changes to their regulatory structures that were not directly in response to the Fukushima Daiichi accident. Two of the countries, France and Sweden, made changes to their structures before the accident occurred. The other two countries (South Korea and the United Kingdom) made changes after the accident but, according to regulatory officials from those countries, these changes were planned prior to the accident. In all four countries, the regulatory changes were consistent with the lessons learned from the Fukushima Daiichi accident.

Specifically, France did not change its regulatory structure in response to Fukushima but overhauled its nuclear regulatory system in 2006 with a law on "Transparency and Security in the Nuclear Field," which established the Autorité de Sûreté Nucléaire (ASN) as an independent commission.[30] ASN has responsibility for nuclear safety, but not physical security, which is the responsibility of a separate authority. French regulatory officials told us that they would prefer that security also fall under the nuclear regulatory body—that integration of these functions would give the regulatory body more expertise and lead to more thoughtful and comprehensive regulation. According to French officials, regulatory and others, the Fukushima Daiichi accident did not drive more nuclear regulatory changes within France because France had already implemented many nuclear safety improvements in response to vulnerabilities exposed in 1999, when strong winds raised the tide, and water breached the seawalls of the Blayais nuclear power plant near Bordeaux and flooded it, including its safety-significant equipment. The plant shut down without incident but, in response, the regulatory body and the plant operator launched a 10-year program to create a more systematic approach to flooding hazards, including raising seawalls and making underground facilities watertight. Studies on flood preparedness at French nuclear power plants were commissioned, and new flood scenarios were considered and modifications made, such as raised seawalls.

Sweden also made changes to its nuclear regulatory structure independently of the Fukushima Daiichi accident. The Swedish Radiation Safety Authority (SSM), established in July 2008 to consolidate various agencies previously involved in nuclear regulation, oversees nuclear safety and security, among other things. In 2010, the Swedish government assigned SSM to review the nation's regulatory framework and strategy. Sweden, like

France, had also already implemented many nuclear safety improvements in response to events prior to Fukushima. For example, the Swedish Parliament decided to require filtered venting equipment to protect against large-scale ground contamination following the Three-Mile Island accident in 1979 in the United States. Sweden also established a Severe Accident Management Program in response to the Three-Mile Island accident.[31]

In 2011, the South Korean government established a new stand-alone regulatory body, the Nuclear Safety and Security Commission (NSSC), which comprises a panel of commissioners that reports directly to the President of South Korea. The establishment of NSSC separated nuclear promotional functions from regulatory ones by removing regulatory decision making from South Korea's Ministry of Education, Science and Technology, which previously oversaw nuclear regulation, as well as nuclear policy and research and development support. South Korean nuclear regulatory officials said that plans for this restructuring were in place before the Fukushima Daiichi accident. According to these officials, NSSC responded to recent nuclear safety incidents, such as the falsification of documents for reactor components at the Yeonggwang nuclear power plant in November 2012, with 20 measures to address such issues, including a program to improve safety culture.[32]

The United Kingdom also created a new regulatory body in April 2011—the Office for Nuclear Regulation (ONR)—to replace the Nuclear Installations Inspectorate, a quasi-autonomous nongovernmental organization. [33] According to the memorandum announcing the body's creation, this change was unrelated to the United Kingdom's response to the Fukushima Daiichi accident. However, the memorandum stated that the events at Fukushima further confirmed the need for a modern, independent and flexible nuclear regulatory body. According to a senior ONR official, the change was prompted by recruitment and retention challenges the nuclear regulatory body was facing in competing with the nuclear industry. According to this official, recruiting from the nuclear industry was a challenge because the pay and conditions within the government, restrained by a public pay policy, were not competitive with those in the industry. An independent study recommended taking the nuclear regulatory body out of the civil service where it would not be subject to the same pay policy. The body remains under the umbrella of the Health and Safety Executive but is not subordinate to it; rather, it is a separate agency. The additional flexibility of this change has allowed ONR to retain staff and to better recruit staff from the industry, which helps meet the needs for new licensing applications. ONR manages the conflicts of interest that could arise from recruiting inspectors directly from industry by starting recruits out in its lower rungs, in technical assessment posts, where inspectors have no contact with operators, with particular care to avoid contact with the operators that had employed them.

# NEARLY HALF OF COUNTRIES WITH NUCLEAR POWER REACTORS IN OUR REVIEW HAVE AUTOMATED DATA TRANSMISSION SYSTEMS

Six of the 13 countries that have operating nuclear power reactors in our review have established automated system for collecting and transmitting critical data to the nuclear regulatory body or designated technical experts who work with the regulatory body during an accident, and a seventh country has plans to build such a system. However, officials from some of these countries, including the United States, told us that their automated data transmission systems are not currently designed to operate under severe emergency conditions, such as loss of off-site power and lines of communication as occurred at the Fukushima Daiichi nuclear power plant.[34] Officials from 5 of the other countries with operating nuclear power reactors told us they either had no plans to establish such a system or they might consider establishing such a system in the future. We were unable to determine whether a sixth country has or plans to establish such a system.

## Six Countries Have Automated Systems, and Some Are Considering Steps to Ensure They Can Operate in Emergency Conditions

Officials from 6 of the 13 countries in our review with operating nuclear power reactors—Japan, the United States, France, South Korea, China, and Argentina—told us that they have automated data transmission systems, and officials in a seventh country, Sweden, told us they are working to install such a system. However, none of these officials told us that their automated data transmission system is currently designed to operate under severe emergency conditions, such as loss of off-site power and lines of communication as occurred at the Fukushima Daiichi nuclear power plant. Although nuclear regulatory officials from 3 countries (Japan, the United States, and France) told us they are considering upgrading their automated data transmission systems to operate under emergency conditions—to a satellite-based system, for example—only Japan, according to a senior Japanese official, has made a final decision to upgrade its system to a more durable and reliable satellite-based communications system. None of these countries' officials told us they have a specific timeline for upgrading their system.

We obtained the most information regarding automated data systems from the United States and France.

### *United States*

NRC established an automated data transmission system in the aftermath of the 1979 Three-Mile Island accident due to a high error rate from plant operators in data transmission during the accident. At that time, NRC relied on telephone calls with its resident inspectors and plant operators to verbally communicate technical data to NRC. In order to substantially improve access to key plant and reactor data, NRC began developing what became the Emergency Response Data System (ERDS) in 1985, and the system became operational in 1993. At present, when activated, ERDS data flow directly to NRC's Operations Center at its Maryland headquarters and are also available, as needed, at NRC regional offices in Pennsylvania, Georgia, Illinois, and Texas. According to NRC officials, approximately 39 of the 62 U.S. nuclear power plants have opted to transmit ERDS data continuously at all times; the remainder has the capability to do so in the event of an incident or emergency. Figure 2 shows NRC's new Headquarters Operations Center.

Source: NRC.

Figure 2. NRC's Headquarters Operations Center.

However, according to NRC officials, ERDS' capacity to collect and transmit information would be compromised under certain emergency conditions, such as the loss of off-site power and lines of communication, as occurred at the Fukushima Daiichi nuclear power plant. Although NRC has not required that ERDS be upgraded to function under these conditions, NRC had taken steps before the Fukushima Daiichi accident to upgrade ERDS— to move the system from an analog based platform to a digital one and to address cybersecurity concerns. In particular, NRC officials told us that, prior to 2009, ERDS used a process requiring the use of a modem to transmit data. Over time, modems have become obsolete, making it difficult to maintain the system. Modems also inherently introduce cybersecurity vulnerabilities. ERDS has been upgraded to use a Virtual Private Network (VPN) to create a secure point-to-point data pathway from each operator site to NRC headquarters.[35] VPN technology, according to NRC, is the current, stable, and reliable information technology industry standard.

NRC is considering additional upgrades to ERDS so that it would function more reliably during severe emergency conditions but has not yet acted because ERDS is not considered a "safety related" system, as it does not directly ensure the safety of a facility or play a direct or essential role in emergency response. These officials added that they thought ERDS was "a nice supplement" to voice communications but that it was not essential because NRC can communicate directly by telephone with its resident inspectors and the operator. They added that telephone communications have been upgraded since the 1979 Three-Mile Island accident. Specifically, NRC funded the establishment of the Emergency Notification System (ENS), which bypasses the local telephone switching network. Plant operators are now required to establish multiple, continuously-staffed ENS telephone lines early in any accident scenario. NRC and licensees have also developed response plans using ENS and conduct frequent exercises to validate these plans.

NRC officials acknowledge that data transmission via ERDS improves the efficiency of information sharing between the licensee and the NRC Headquarters Operations Center but said that the practices and procedures used to obtain the data using ENS would enable NRC to evaluate the severity of the reactor accident without ERDS if necessary. Accordingly, these officials told us that NRC is first completing higher priority post-Fukushima nuclear safety enhancements before deciding whether or how to upgrade ERDS because how these enhancements are resolved could affect how any ERDS upgrade might be implemented. They also noted that both the Three-Mile Island and Fukushima Daiichi accidents were managed without the benefit of functioning automated data transmission systems. However, an IAEA fact-finding mission to Japan in the aftermath of Fukushima stated that greater consideration should be given in the future to providing hardened systems, communications, and sources of monitoring equipment for providing essential information for on-site and off-site responses, especially for severe accidents. Moreover, without upgrades to enable ERDS to function under emergency conditions, it may not be available when the need for it is greatest, such as during a severe accident.

### *France*

French officials told us, during our tour of one of their national crisis centers—the French government's emergency response center for dealing with events with possible radiological implications—that France also does not consider its automated data transmission system a "safety system." These officials told us that there are accident scenarios under which emergency diesel generators would power their automated data transmission system, but whether the system would function would depend on the severity of the accident and level of damage to the plant's electrical system. In particular, if power were lost during a severe accident as it was during the Fukushima Daiichi accident, the automated system would not function. French officials also told us that they maintain close telephone contact with plant operators during an accident. Regardless of whether its automated data transmission system functions, France requires that "quarter-hour messages" be sent from the nuclear power plant control room to technical experts at its national crisis centers. According to French officials, these messages convey data on the most important nuclear power reactor parameters, a subset of data collected by the automated transmission system. These data are read by plant operator technicians from the control room instruments, copied by hand to a paper form, and then faxed to its national crisis centers every 15 minutes. Nonetheless, France is also considering upgrading its automated data transmission system to reliably function without off-site power; according to French officials, studies are in progress, but no decisions have been made.

Nuclear regulatory officials we spoke with from 5 of the 6 remaining countries with operating reactors—Armenia, Belgium, Canada, Russia, and the United Kingdom—told us they do not currently have automated data transmission systems, and officials from 3 of these countries said their countries may consider establishing such systems in the future. We were unable to determine whether a sixth country, Pakistan, has or plans to establish such a system.

For the 3 countries with emerging or potential nuclear power programs, regulatory officials in Vietnam told us that they plan to develop an automated data transmission system; officials in the United Arab Emirates told us they have not yet decided whether to do so and, in the case of Indonesia, which has not yet decided whether to pursue a civilian nuclear power program, we were unable to obtain information.

## International Efforts Taken to Support Nuclear Regulatory Bodies and Promote Nuclear Safety

Three key international organizations—IAEA, WANO, and the EU—along with the Convention on Nuclear Safety have taken steps to support nuclear regulatory bodies and promote nuclear safety worldwide since the Fukushima Daiichi accident—including playing a significant role in helping nuclear regulatory bodies identify the most important lessons from the accident and supporting the associated regulatory changes to enhance nuclear safety.

### IAEA

IAEA adopted an Action Plan on Nuclear Safety (Action Plan) in September 2011, 6 months after the Fukushima Daiichi accident. The plan's purpose is to define a program of work to strengthen global nuclear safety by taking into account lessons learned from the accident. The Action Plan focuses on, among other things, safety assessments of nuclear power plants in light of their vulnerabilities to extreme natural hazards, peer reviews of nuclear regulatory bodies and nuclear power plants, provisions for emergency preparedness and response, and the adequacy of existing IAEA safety standards. The success of the Action Plan, according to IAEA, depends on its implementation through the full participation and cooperation of the member states, with the involvement of relevant stakeholders such as international organizations, nuclear industry associations, and research organizations. The plan is divided into 12 actions, each of which is divided into subactions and activities. The status of each action is tracked at the activity level and posted on IAEA's publicly available website. According to IAEA officials, the agency does not track the efforts of member states under the Action Plan but encourages member states to share information on progress in the implementation of the Action Plan through international conferences and international experts' meetings on nuclear safety-related topics.

One of the key ways in which IAEA helps countries improve nuclear safety and regulatory effectiveness is through peer review missions. The most prominent peer review missions are the Integrated Regulatory Review Service (IRRS) and Operational Safety Review Team (OSART) missions. These missions evaluate the operations of a member state's nuclear regulatory system and civilian nuclear power plant operational safety, respectively, and make recommendations and suggestions for improvement. According to IAEA officials, countries and plant operators hosting peer review missions are also encouraged and expected by IAEA to host follow-up missions, typically 2 to 4 years after the initial review in the case of IRRS missions and 18 to 24 months in the case of OSART missions. According to these officials, follow-up missions have been a part of IRRS and OSART peer reviews since 2006 and 1989, respectively.

IRRS missions, which are voluntary to all but EU member states, assess the safety practices of the requesting country through an examination of its regulatory framework and organization and compare the country's practices with IAEA Safety Standards and good regulatory practices.[36] IRRS mission teams include experts drawn largely from IAEA member countries' nuclear regulatory bodies, including NRC. The requesting country is typically provided with recommendations and suggestions for improvement at the conclusion of the mission. According to IAEA officials, recommendations are proposed to host countries where aspects of their regulatory system relative to the IAEA Safety Requirements are missing, incomplete, or inadequately implemented. These recommendations, according to IAEA officials, should be specific, realistic, and designed to result in tangible improvements to regulatory effectiveness. When IAEA peer reviewers identify opportunities for improvement not directly related to inadequate conformance with IAEA Safety Requirements, but that may contribute to improvements in national regulatory arrangements, they make IRRS suggestions. According to IAEA, suggestions are primarily intended to make the regulatory body's performance more effective or efficient, and to point out potential improvements to current regulatory activities.

IAEA conducted 75 IRRS missions to 45 countries from 1992 through the end of June 2013. [37] In addition, according to IAEA officials, a special module focused on the regulatory implications of the Fukushima Daiichi accident was incorporated into the scope of IRRS missions immediately after the accident for countries having nuclear power plants. Table 1 lists the number of IRRS missions, including follow-up missions, which member states have hosted through June 30, 2013.

**Table 1. Number of IAEA Integrated Regulatory Review Service (IRRS) Missions by Country, 1992 through June 30, 2013**

| Country | Number of IRRS missions |
|---|---|
| Armenia | 2 |
| Australia | 2 |
| Botswana | 1 |
| Bulgaria | 3 |
| Cameroon | 1 |
| Canada | 2 |
| China | 3 |
| Côte d'Ivoire | 1 |
| Czech Republic | 2 |
| Finland | 3 |
| France | 2 |
| Gabon | 1 |
| Germany | 2 |
| Greece | 1 |
| Hungary | 2 |
| Indonesia | 2 |
| Iran | 1 |
| Japan | 1 |
| Kenya | 1 |
| Korea, Republic of | 1 |
| Lebanon | 1 |
| Lithuania | 1 |
| Madagascar | 1 |
| Mauritius | 1 |
| Mexico | 2 |
| Namibia | 1 |
| Niger | 1 |
| Pakistan | 1 |
| Peru | 1 |
| Poland | 1 |
| Romania | 5 |

| Country | Number of IRRS missions |
|---|---|
| Russian Federation | 1 |
| Sierra Leone | 1 |
| Slovakia | 3 |
| Slovenia | 2 |
| Spain | 2 |
| Sweden | 1 |
| Switzerland | 3 |
| Thailand | 1 |
| Uganda | 1 |
| Ukraine | 4 |
| United Arab Emirates | 1 |
| United Kingdom | 2 |
| United States | 1 |
| Vietnam | 2 |
| Total | 75 |

Source: GAO analysis of IAEA data.

Note: IRRS missions were preceded by a similar program from 1992 to 2004 called International Regulatory Review Team (IRRT) missions. This table combines both types of missions. IAEA does not consider scoping or expert missions to be "full" IRRT or IRRS missions and are thus not included in this table.

OSART missions also include experts drawn largely from IAEA member countries but are more narrowly targeted to review operational safety at specific nuclear power plants. OSART missions also typically make recommendations and suggestions to improve safety. According to IAEA, OSART recommendations provide operators ideas for improving operational safety and should: (1) be based on IAEA Safety Standards or proven, good international practices; (2) address the root causes rather than the symptoms of the identified concern; and (3) be specific, realistic, and designed to result in tangible improvements. An OSART suggestion is primarily intended to, among other things, make good performance more effective. In general, suggestions are designed to stimulate the plant management and supporting staff to continue to consider ways and means for enhancing performance. According to IAEA, the agency has conducted 284 OSART missions in 34 countries from 1983 through the end of June 2013. The Fukushima Daiichi nuclear power station, like most nuclear power plants, has not hosted an OSART mission. Table 2 lists the number of OSART missions, including follow-up missions, that member countries have hosted through June 2013.

**Table 2. Number of IAEA Operational Safety Review Team (OSART) Missions by Country, 1983 through June 30, 2013**

| Country | Number of OSART missions |
|---|---|
| Argentina | 2 |
| Armenia | 2 |
| Belgium | 4 |
| Brazil | 11 |
| Bulgaria | 8 |
| Canada | 4 |
| China | 19 |
| Czech Republic | 15 |
| Finland | 4 |
| France | 44 |
| Germany | 9 |
| Hungary | 4 |
| India | 1 |
| Italy | 3 |
| Japan | 9 |
| Kazakhstan | 1 |
| Korea, Republic of | 8 |
| Lithuania | 4 |
| Mexico | 6 |
| Netherlands | 5 |
| Pakistan | 6 |
| Philippines | 2 |
| Poland | 1 |
| Romania | 6 |
| Russian Federation | 12 |
| Slovakia | 10 |
| Slovenia | 6 |
| South Africa | 6 |
| Spain | 9 |
| Sweden | 13 |
| Switzerland | 9 |
| Ukraine | 23 |
| United Kingdom | 6 |
| United States | 12 |
| Total | 284 |

Source: GAO analysis of IAEA data.

According to IAEA officials, the agency is not systematically tracking the status of the recommendations made by the IRRS or OSART peer review missions, and it does not know the extent to which the recommendations or suggestions have been implemented by the host countries and plant operators unless countries and operators have hosted a follow-up mission. Follow-up missions to IRRS and OSART missions are the means for informing IAEA of progress made in implementing the original missions' recommendations and suggestions. IAEA posts on its publicly available website the names of countries and nuclear power plants that have hosted IRRS and OSART missions and follow-up missions and, in some cases, summary or full reports of the main conclusions of a given mission.[38] According to IAEA officials, however, the agency is not authorized by member states to act as a nuclear safety oversight organization and so does not systematically track the status of such recommendations and suggestions. IAEA officials noted that member states are responsible for tracking the status of their mission findings. However, without this information, IAEA cannot determine the impact and effectiveness of the peer review missions. Nuclear regulatory officials from France and Canada stated that greater transparency about the results and follow-up to peer review missions is needed.

The importance of tracking recommendations and improving the impact and effectiveness of IRRS missions was noted at a recent IAEA conference. Specifically, the first action item at the April 2013 IAEA regulators conference in Ottawa, Canada, identified issues that needed to be addressed, implemented, and followed up on, including: (1) countries should develop national action plans to implement recommendations made by IRRS peer review missions, (2) the results of the IRRS missions should be made public, and (3) IRRS missions and follow-up missions should be reported to IAEA's Board of Governors. Because not all of the recommendations and suggestions of IRRS peer review missions are made public, countries may have less incentive to follow up on them.

The experience of the 2011 IAEA-sponsored international expert mission to Japan highlights the current system's limitations. This expert mission, which was established to find facts and identify initial lessons from the Fukushima Daiichi accident, noted, among other things, that despite the technical focus on nuclear safety and radiation protection, the mission also identified more general problems with: the regulatory framework for safety in Japan, the clarity of the roles and responsibilities among government entities and the operator, and the independence of the nuclear regulatory body. The expert mission noted that some of these issues had also been raised by the June

2007 IRRS mission to Japan, and that there had not been a follow-up mission. The IRRS report did not mention "regulatory capture"— a condition that was found by the Kurokawa Commission to have compromised Japan's nuclear safety prior to the accident.[39]

## WANO

In the wake of the Fukushima Daiichi accident, WANO is expanding the scope of its operations and activities to do more to help its members achieve its overriding goals to enhance nuclear safety and the operational performance of nuclear power plants. Specifically, according to WANO, the organization is

- *Increasing staff.* WANO is more than doubling the size of its staff over 3 years. It had about 170 employees in 2011 and plans to increase the number to 388 in 2014.
- *Requiring more frequent peer reviews focusing on issues highlighted by the accident.* WANO is moving from 6-year to 4-year intervals for the peer reviews it requires of its members. WANO is also requiring its members to host a WANO peer review team for a follow-up 2 years after its peer reviews so the operator can explain what it has done to address the findings from the previous review. In addition, WANO has now expanded the scope of its peer reviews to include, among other things, issues highlighted by the accident, including emergency preparedness, severe accident prevention, and spent fuel pool management.
- *Increasing coordination with IAEA.* WANO also signed a Memorandum of Understanding with IAEA in September 2012 under which both organizations are to take a more coordinated approach to their activities. For example, under this memorandum, according to both organizations, WANO and IAEA staff are to meet regularly to discuss major safety related activities, share more nuclear power plant operating experience, and better coordinate the timing of their respective peer reviews so that, for example, operators will not have to host a WANO peer review in the same year they host an IAEA OSART mission.

## European Union

The EU, acting in concert with national nuclear regulatory bodies and the nuclear industry, among its member states, launched a process in March 2011 to carry out EU-wide comprehensive risk and safety assessments of nuclear power plants. The EU nuclear safety directive and the EU nuclear waste directive grant it the authority to make these assessments legally binding on EU members. Within a few weeks of the Fukushima Daiichi accident, all 14 EU member states that operate civilian nuclear power reactors—as well as two neighboring countries—had agreed, on a voluntary basis, to subject their power plants to targeted, comprehensive safety risk assessments of how well they could withstand the consequences of various extreme external conditions, including those that affected the Fukushima Daiichi plant.[40] These assessments were to focus on the robustness of safety margins of nuclear power plants in three areas: (1) accident-initiating events such as earthquakes, floods, and extreme natural weather; (2) loss of safety systems, including power and ultimate heat sink;[41] and (3) severe accident management capabilities. The methodology of these assessments was determined by EU experts from the European Nuclear Safety Regulators Group (ENSREG) and the West European Nuclear Regulators' Association (WENRA).[42] The comprehensive safety and risk assessments took place from June 2011 through April 2012 and were conducted with a three-step process as follows:

1) Assessment of plants by operators (June to October 2011).
2) Review of these assessments by national nuclear regulatory bodies (by the end of December 2011).
3) European peer reviews of assessments (January to April 2012).

Following the completion of these safety assessments, according to EU officials, the ENSREG action plan was approved on August 1, 2012. The plan is to, among other things, (1) assist in ensuring that the conclusions from the comprehensive safety and risk assessments and their peer review result in improvements in safety across European nuclear power plants and (2) ensure, through further peer review, that the recommendations and suggestions from the peer reviews of comprehensive safety and risk assessments (also known as "stress tests") are addressed by national regulators and ENSREG in a consistent manner.

The EU also undertook a review of the existing legal and regulatory frameworks for the safety of nuclear installations to consider proposing any improvements that may be necessary. Having carried out EU-wide comprehensive risk and safety assessments of nuclear power plants, as well as a review of the existing legal and regulatory frameworks for the safety of nuclear installations, the European Commission concluded that improvements to European nuclear safety were necessary. The commission announced in June 2013 its proposal to update and strengthen its 2009 Nuclear Safety Directive. According to the commission announcement, the proposed new directive substantially strengthens the provisions of the existing directive by

- introducing EU-wide safety objectives;
- setting up a European system of peer reviews of nuclear installations;
- increasing transparency on nuclear safety matters;
- strengthening the role and independence of national regulatory authorities;
- introducing a requirement of specific safety reviews for older nuclear power plants for which a lifetime extension is considered; and
- enhancing on-site emergency preparedness and response, for example by implementing strict accident management guidelines and by putting in place emergency response centers that must be protected against radioactivity and earthquakes or flooding.

If ultimately adopted by the European Council, this proposed new directive could enter into force as soon as 2014 and be written into binding national legislation by EU member states within the following 18 months.

## Convention on Nuclear Safety

The Convention has convened two meetings since the Fukushima Daiichi accident—the regularly scheduled triennial review meeting in April 2011, which was held, coincidentally, about 3 weeks after the accident, and a special meeting approximately 18 months later, in August 2012, to focus specifically on the accident. For the special August 2012 meeting, parties to the Convention (1) prepared a Fukushima Daiichi-related national report describing the safety assessments of their nuclear programs undertaken after the accident and measures taken to address the lessons learned from the accident; (2) reviewed and shared their findings and measures taken with other parties through the peer review process and, to varying extents, the public; (3) reviewed the effectiveness and provisions of the Convention.

One result of these meetings was to make and adopt revisions to the Convention's guidelines on the preparation of the triennial national reports.[43] The revised post-Fukushima guidelines provide more detailed suggestions for what information parties should include in their triennial national report and how it should be organized. Parties are encouraged to

- refer to IAEA safety standards, such as IAEA Safety Fundamentals and Requirements, in discussing how they are meeting their obligations to the Convention;
- report in more detail about the steps they have taken to follow up on any issues and concerns that may have been raised in earlier review meetings and national reports;
- discuss and address the results of international peer review missions, including IAEA IRRS missions, progress made in implementing recommendations, and suggestions and additional plans for further follow-up; and
- discuss lessons they have learned or actions taken in response to accidents and events, such as the issues highlighted by the Fukushima Daiichi accident.

The parties also agreed to consider other measures to improve the effectiveness of the Convention at the next review meeting, scheduled for March 2014. In particular, part of the outcome of the August 2012 special meeting was the establishment of an effectiveness and transparency working group open to all interested parties to the Convention, to develop and propose various actions that could be taken to strengthen and amend, where necessary, the Convention. This working group is to consider the various proposals raised during the special meeting to amend the Convention, including proposals already submitted by Switzerland and the Russian Federation. The working group is to present its findings to all parties at the 2014 review meeting. Among the Swiss proposals are several amendments to the Convention, including one that would require each contracting party to have periodic external review of its regulatory body to confirm its competence, independence, and that it is provided with adequate authority and resources. Another Swiss proposal would also require that the regulatory body's findings and decisions on the safety of nuclear installations be made publicly available. The Russian proposals include (1) requiring periodic assessments of nuclear power plant safety and (2) affirming that new construction conforms with IAEA recommendations. NRC officials noted that the U.S. position is that changes to the Convention could be better addressed through guidance documents, rather than through reopening the Convention to amendments.

As we reported in April 2010, IAEA officials told us it was important for parties to make as much information about their civilian nuclear power programs accessible as possible but that it was for each party to determine how much information should be made public and how much should remain confidential.[44] We also recommended in 2010 that the Secretary of State, in coordination with the NRC Chairman, work with other parties to the Convention to, among other things, expand efforts to increase the number of parties' national reports made available to the public by posting them to IAEA's public website, even though there is no requirement to do so. In response to these recommendations, NRC stated that it (1) leads by example by always making the U.S. national report available on the NRC and IAEA publicly available websites and (2) uses its leadership positions in the Convention to work with and encourage leaders of other countries to encourage all contracting parties to make their national reports publicly available. In addition, under U.S. leadership, according to NRC, the Convention on Nuclear Safety Scientific Secretary in 2010 sent out a message to all contracting parties reminding them that "in the spirit of openness and transparency, Contracting Parties are encouraged to notify IAEA when they agree to have their National Reports posted on the IAEA public website." Further, NRC and the State Department stated that they will also raise this issue during meetings of the Convention working group on effectiveness and transparency and continue to encourage contracting parties to make as much information publicly accessible as possible.

Since April 2010, an additional 13 national reports prepared for the four review meetings held from 1999 to 2008 have been made available on IAEA's publicly available website, including 8 additional national reports prepared for the 2008 review meeting. Notwithstanding this overall increase in the number of reports posted, the total number of reports posted on IAEA's publicly available website for the 2011 review meeting declined to 32 reports from 34 reports in 2008 despite the increase in the number of contracting parties to the Convention obligated to submit national reports increasing from 61 to 72 parties. Moreover, according to summary reports prepared by the parties at the end of the August 2012 special meeting called to discuss the Fukushima Daiichi accident, 36 of 61 parties (or 59 percent) that prepared national reports for the meeting did not make them available on IAEA's public website.[45]

## CONCLUSION

The Fukushima Daiichi accident underscored the importance of countries having a strong, competent, and independent regulatory body and the consequences of not having one. Nuclear regulatory bodies worldwide are taking this accident seriously and using it as an opportunity to strengthen regulatory procedures, requirements, and infrastructures, as appropriate. Nuclear regulatory bodies will continue to play key roles in minimizing the chances of a severe accident and ensuring the most effective emergency response possible should an accident occur— factors considered to be fundamental to public safety and to the continued viability of the civilian nuclear power industry worldwide. For example, NRC participates regularly in a variety of IAEA activities related to, among other things, nuclear safety, and the State Department coordinates U.S. policy matters with international organizations and treaties, including those dealing with nuclear safety. Supporting these efforts, IAEA peer review missions provide countries with unique opportunities to strengthen their nuclear regulatory organizations. However, IAEA does not systematically track the status of the recommendations made by the IRRS peer review missions and does not know the extent to which the recommendations have been implemented by the host countries. According to IAEA officials, the agency is not authorized by member states to act as a nuclear safety oversight organization, and member states are responsible for tracking the status of their mission findings. Without this information, IAEA cannot fully determine the impact and effectiveness of these peer review missions as important tools to improve nuclear safety. Furthermore, because not all of the recommendations and suggestions of the IRRS peer review missions are made public, countries may have less incentive to follow up on them.

Although NRC does not consider its automated data transmittal system—ERDS—a safety system, the Fukushima Daiichi accident underscored the usefulness of having such a system fully functioning. NRC has taken steps to upgrade ERDS, but it is still not equipped to collect and transmit information when electrical power and lines of communication are compromised, as occurred at the Fukushima Daiichi nuclear power plant. NRC is delaying its consideration of additional upgrades to ERDS so that it would function more reliably because it is first completing higher priority post-Fukushima nuclear safety enhancements. By delaying its decision on whether or how to upgrade ERDS, a fully functioning system might not be available to regulators when the need is greatest—during a severe accident. Further, NRC may be missing an opportunity to lead by example as it does in other areas to promote nuclear safety worldwide, such as its efforts to encourage parties to the Convention on Nuclear Safety to make their national reports publicly available.

## RECOMMENDATIONS FOR EXECUTIVE ACTION

We are making the following two recommendations in this report:

- To further promote the safety of civilian nuclear power programs worldwide by enhancing the effectiveness of nuclear regulatory bodies, we recommend that the Secretary of State, in coordination with the Commission, work with and encourage officials from IAEA to systematically track the status of the recommendations made by the IRRS peer review missions and make this information publicly available to the extent feasible.
- To increase the likelihood of NRC's access to timely, accurate, and comprehensive information during nuclear accidents, we recommend that the NRC Chairman consider expediting NRC's decision on whether or how to upgrade ERDS so that it would remain functional during a severe accident.

## AGENCY COMMENTS AND OUR EVALUATION

We provided a draft of this report to the Department of Energy, the Department of State, and NRC for review and comment. We also provided relevant sections of the draft report to IAEA. NRC stated that it was in general agreement with the draft report but did not state whether it agreed or disagreed with our two recommendations, including the one recommending that NRC consider expediting its decision on whether or how to upgrade ERDS so that it would remain functional during a severe accident. The Department of Energy provided technical comments, as did IAEA and NRC, which we incorporated as appropriate.

In its written comments, the Department of State "partially concurred" with our recommendation that it, in coordination with the NRC Chairman, work with officials from IAEA to systematically track the status of the recommendations by the IRRS peer review missions and make this information publicly available to the extent feasible. In its comment letter, the department stated that it would work with NRC and raise the recommendation with appropriate safety officials at IAEA and seek their views to further discuss what possible steps could be taken. However, State raised a number of issues that it believed would need to be considered in discussions with IAEA officials, including: (1) maintaining member states' confidentiality and willingness to participate in IRRS missions if IAEA systematically tracks the status of peer review recommendations and makes those results public, (2) member states' acceptance of such tracking without extensive prior discussions, and (3) IAEA's perspective on whether such tracking would improve the focus of the missions. State also raised questions about the costs associated with such systematic tracking and how they will be paid. We are encouraged that the department plans to work with NRC and raise the recommendation with IAEA and hope that these issues can be addressed in discussions with IAEA officials. NRC echoed issues similar to State's in its technical comments and also noted that its more appropriate role should be to encourage—rather than work with—IAEA to systematically track the status of the recommendations by the IRRS peer review missions. In response to NRC's comment, we have added the word "encourage" to the recommendation so that it now reads, in part, that "the Secretary of State, in coordination with the Commission, work with and encourage officials from IAEA to systematically track the status of the recommendations by the IRRS peer review missions . .."

We recognize the significance of the issues State and NRC have raised about maintaining confidentiality, member states' acceptance of systematic tracking, how such tracking might affect member states' willingness to participate in future IRRS missions, and the focus of such missions. We agree that these issues need to be factored into discussions with IAEA officials. For this reason, we recommended that the results of these missions be made publicly available to the extent feasible. If discussions among State, NRC, and IAEA officials can lead to greater transparency about the results of IRRS missions and improve prospects for analyzing the impact and effectiveness of these missions, they would be worthwhile. However, as we noted in our report, without systematic tracking of the results of mission findings, including follow-up, IAEA cannot fully determine the impact and effectiveness of the peer review missions.

In addition, our recommendation is consistent with another ongoing international effort—with U.S. participation—to encourage countries to be more open regarding nuclear safety.

For example, as we noted in our report, the most recent version of the Convention on Nuclear Safety's guidelines encouraged the contracting parties to discuss and address the results of international peer review missions in their national reports, including IRRS missions; discuss progress made in implementing recommendations; and discuss suggestions and additional plans for further follow-up.

Regarding State and NRC's mention of potential added costs associated with any tracking system, we note that neither agency provided any analysis or assessment of the potential added costs of systematic tracking in raising it as an issue.

Sincerely yours,

David C. Trimble
Director,
Natural Resources and Environment

## APPENDIX I: SCOPE AND METHODOLOGY

In light of the Fukushima Daiichi accident, you asked us to examine worldwide nuclear regulation and safety. This report examines: (1) the actions regulatory agencies with existing or planned civilian nuclear power reactors have taken to strengthen nuclear safety; (2) the extent to which these countries have established automated systems for collecting and transmitting data and taken steps to enable such systems to withstand emergency conditions; and (3) steps key international organizations have taken to support nuclear regulatory bodies and promote nuclear safety worldwide since the Fukushima accident and the impact of some of these steps.

To examine the actions that national regulatory agencies have taken to strengthen nuclear safety in the aftermath of the Fukushima accident, we selected a nonprobability sample of 16 countries to represent a cross section of "emerging" and established civilian nuclear power countries. For countries with emerging civilian nuclear power, we selected the three most often suggested by U.S. agency officials and nuclear experts as available. For established countries, we focused on those where nuclear power represented more than 15 percent of the country's electricity production. From those, we asked agency officials and other interviewees to suggest countries that would provide good examples of actions taken to strengthen nuclear safety.

We also considered program size and the nation's nuclear regulatory framework to select 13 countries with established civilian nuclear power.[1] From the sample, we prioritized 5 countries to visit in person. We selected these 5 countries through a combination of suggestions from experts, size of nuclear program, and proximity to one another. We interviewed, at an international nuclear regulators' conference in Ottawa, Canada, regulatory officials from all but 1 of the 11 countries we did not visit in person; Armenia, the country with regulatory officials we did not interview in person, provided written answers to our questions.

The 16 countries are Argentina, Armenia, Belgium, Canada, China, France, Indonesia, Japan, Pakistan, Russia, South Korea, Sweden, United Arab Emirates, United Kingdom, United States, and Vietnam. Thirteen of these 16 countries—those with established nuclear power programs—operated 78 percent of the world's nuclear power reactors as of September 2013 (the other 3 countries—Indonesia, United Arab Emirates, and Vietnam—do not have operating nuclear power reactors).[2] For the 16 countries selected, we reviewed reports these countries undertook to document the lessons learned from the Fukushima accident, including reports prepared for various international meetings—including the Second Extraordinary Meeting for the Convention on Nuclear Safety—and reports relating to their comprehensive safety risk assessments, and/or presentations in which they summarized lessons learned.[3] We also interviewed officials from each selected regulatory body and/or technical support organization and/or received written responses to our questions.

In countries where we conducted in-person site visits—Belgium, Canada, France, Sweden, and the United States— we also visited nuclear power plants and interviewed nuclear power plant operators. In Canada, France, Sweden, and the United States, we interviewed other government officials in agencies or ministries related to civilian nuclear power. In the Canada, the United Kingdom, and the United States, we interviewed nuclear industry association officials. We also interviewed two former chairmen of the Nuclear Regulatory Commission (NRC). GAO did not independently verify statements of foreign law.

We selected six illustrative categories of actions taken in response to Fukushima; our selection was neither an attempt to identify the full range of regulatory responses to the Fukushima Daiichi accident, nor to establish them as best practices or to catalog the extent to which each regulatory body within our sample undertook actions in each category.

To examine the extent to which the 15 selected non-U.S. countries have established automated systems for collecting and transmitting data to the nuclear regulatory body during emergencies, we interviewed officials from the respective nuclear regulatory agencies, technical support organizations, and operators, as applicable, and possible and corroborated these statements with documents to the extent possible. GAO did not receive documentary or other substantial corroborating information to substantiate the statements of officials from China and Argentina that they have automated data transmission systems, and we have received only limited corroborating information that South Korea has such a system. There is no requirement that officials from these countries provide such information.

To examine the steps key international organizations have taken to support nuclear regulatory bodies and promote nuclear safety worldwide since the Fukushima Daiichi accident, we first identified such organizations through interviews with officials from the State Department, Department of Energy (DOE), and nuclear regulatory officials. Based on these discussions, we identified the International Atomic Energy Agency (IAEA), the Organization for Economic Cooperation and Development's (OECD) Nuclear Energy Agency (NEA), the World Association of Nuclear Operators (WANO), and the European Union (EU). We then reviewed, as applicable, the reports and action plans and other documents from these organizations and interviewed or exchanged responses to written questions with relevant officials from them. We also reviewed NRC documents and responses to written questions from NRC officials, and we reviewed State Department documents. In the case of IAEA, we reviewed IAEA information and details about IRRS and OSART missions, as well as answers to written questions and conference documents. We also reviewed previous GAO work and statements from IAEA officials and other officials from the Ottawa conference.

We conducted this performance audit from May 2012 to March 2014 in accordance with generally accepted government auditing standards. Those standards require that we plan and perform the audit to obtain sufficient, appropriate evidence to provide a reasonable basis for our findings and conclusions based on our audit objectives. We believe that the evidence obtained provides a reasonable basis for our findings and conclusions based on our audit objectives.

# APPENDIX II: CIVILIAN NUCLEAR PROFILES OF SELECTED COUNTRIES

| Country | Nuclear regulatory agency | Established[a] | Structure | Scope of regulatory authority[b] | Operational nuclear reactor and electricity generation data as of September 2013 |
|---|---|---|---|---|---|
| Argentina | Argentine Nuclear Regulatory Authority (ARN) | 1997 | Autonomous entity within the jurisdiction of the General Secretariat of the Presidency (reports to the President of Argentina). | Nuclear safety; nuclear security; nuclear nonproliferation; environment and public health. | 2 power reactors (and 1 under construction) generating 5% of the country's electricity |
| Armenia | Armenian Nuclear Regulatory Authority (ANRA) | 1993 | The ANRA Chairman is appointed and dismissed by the Prime Minister and reports to the President, government, and Prime Minister, but not to any other ministry. | Nuclear safety; nuclear security; nuclear nonproliferation; environment and public health; nuclear waste. | 1 power reactor generating27% of the country's electricity |
| Belgium | Federal Agency for Nuclear Control (FANC) | 1994 | An autonomous public institution that reports to the Minister of Home Affairs. | Nuclear safety; nuclear security; nuclear nonproliferation; environment and public health; nuclear waste. | 7 power reactors generating 51% of the country's electricity |
| Canada | Canadian Nuclear Safety Commission (CNSC) | 2000 | The Commission Tribunal, CNSC's central decision-making body, is a quasi-judicial administrative tribunal. | Nuclear safety; nuclear security; nuclear nonproliferation; environment and public health; nuclear waste. | 19 power reactors generating 15% of the country's electricity |

## Appendix II. (Continued)

| Country | Nuclear regulatory agency | Established[a] | Structure | Scope of regulatory authority[b] | Operational nuclear reactor and electricity generation data as of September 2013 |
|---|---|---|---|---|---|
| China | National Nuclear Safety Administration (NNSA) | 1984 | Under the Ministry of Environmental Protection. | Nuclear safety; environment and public health; nuclear waste. | 18 power reactors (and 28 under construction) generating 2% of the country's electricity |
| France | Autorité de Sûreté Nucléaire (ASN) | 2006 | ASN's 5-person Commission serve for a 6-year, full-time term. ASN reports to the French Parliament. | Nuclear safety; nuclear security;[c] environment and public health; nuclear waste. | 58 power reactors (and 1 under construction) generating 75% of the country's electricity |
| Indonesia | BAPETEN | 1997 | BAPETEN has the power to establish regulations and to conduct licensing and inspections. BAPETEN is under and directly responsible to the President of Indonesia. | Nuclear safety; nuclear security; nuclear nonproliferation; environment and public health; nuclear waste. | No power reactors (3 research reactors, regulated by BAPETEN) |
| Japan | Nuclear Regulatory Authority (NRA) | 2012 | An independent commission (under the Ministry of the Environment). | Nuclear safety; nuclear security; nuclear nonproliferation; environment and public health; nuclear waste. | 50 power reactors (and 2 under construction) generating 2% of the country's electricity[d] |
| Korea, Republic of | Nuclear Safety & Security Commission (NSSC) | 2011 | The 9-person commission an independent, stand-alone, ministry-level agency and supported by two technical support organizations. | Nuclear safety; nuclear security, nuclear nonproliferation; environment and public health; nuclear waste. | 23 power reactors (and 5 under construction) generating 30% of the country's electricity |

| **Country** | **Nuclear regulatory agency** | **Established**[a] | **Structure** | **Scope of regulatory authority**[b] | **Operational nuclear reactor and electricity generation data as of September 2013** |
|---|---|---|---|---|---|
| Pakistan | Pakistan Nuclear Regulatory Authority (PNRA) | 2001 | PNRA is an independent regulator reporting directly to the Prime Minister. PNRA comprises a chairman and 2 full-time and 7 part-time members. The federal government appoints the chairman and the members. | Nuclear safety; nuclear security; environment and public health; nuclear waste. | 3 power reactors (and 2 under construction) generating 5% of the country's electricity |
| Russian Federation | Rostechnadzor (the Federal Environmental, Industrial and Nuclear Supervision Service) | 2004 | Rostechnadzor is directly subordinate to the government and not to any ministry. | Nuclear safety; nuclear security; environment and public health; nuclear waste. | 33 power reactors (and 10under construction) generating 18% of the country's electricity |
| Sweden | The Swedish Radiation Safety Authority (SSM) | 2008 | SSM reports to the Ministry of the Environment but interprets and executes the law independently of the government. | Nuclear safety; nuclear security; nuclear non-proliferation; environment and public health; nuclear waste. | 10 power reactors generating 38% of the country's electricity |
| United Arab Emirates | Federal Authority for Nuclear Regulation (FANR) | 2009 | FANR is an independent legal entity with "full legal competence and financial and administrative independence." It is fully separate from the nuclear sector. | Nuclear safety; nuclear security; nuclear nonproliferation nuclear waste. | 2 power reactors under construction |

# Appendix II. (Continued)

| Country | Nuclear regulatory agency | Established[a] | Structure | Scope of regulatory authority[b] | Operational nuclear reactor and electricity generation data as of September 2013 |
|---|---|---|---|---|---|
| United Kingdom | Office for Nuclear Regulation (ONR) | 2011 | ONR is a stand-alone statutory body, formed as an agency of the Health and Safety Executive but working toward becoming an independent corporation. | Nuclear safety; nuclear security; nuclear nonproliferation. | 16 power reactors generating 18% of the country's electricity |
| United States of America | Nuclear Regulatory Commission (NRC) | 1974 | Independent federal agency headed by five commissioners (commission is a quasi-judicial body). | Nuclear safety; nuclear security; nuclear nonproliferation; nuclear waste. | 100 power reactors (with 3 under construction) generating 19% of the country's electricity |
| Vietnam | Vietnam Agency for Radiation and Nuclear Safety (VARANS) | 2003 | VARANS is under the Ministry of Science and Technology. | Nuclear safety; nuclear security; nuclear nonproliferation. | No power reactors (1 research reactor, regulatedby VARANS) |

Source: GAO analysis of nuclear regulatory body documents; IAEA Power Reactor Information System as of September 4, 2013.

Note: Scope also implies only civilian and commercial nuclear matters; this table does not specify the regulatory agencies that also have oversight over military nuclear matters.

[a]This date represents the establishment of each country's current nuclear regulatory body, which may have replaced an existing nuclear regulatory body.

[b]This column represents each regulatory body's scope of responsibilities as reported by the country and does not convey any GAO analysis of the regulatory body's performance of its responsibilities. Scope of regulatory authority comprises areas addressed by technical support organizations, such as Belgium's Bel-V, France's Institut de Radioprotection et de Sûreté Nucléaire (IRSN), and Korea's KINS; areas not included may be performed by government agencies. For example, the U.S. NRC is not the primary U.S.

agency responsible for oversight of environment and public health, but shares in these responsibilities with regard to civilian nuclear power with other federal agencies and state and local governments.

[c]France's ASN is responsible for oversight of nuclear security (for example, security of radioactive materials, including transportation of such materials) but not for oversight of physical protection of nuclear power plants. However, its technical support organization, IRSN, has a role in monitoring physical protection.

[d]Before all but two of Japan's remaining 50 reactors were ordered to suspend operations after the Fukushima-Daiichi accident, nuclear reactors generated 18% of Japan's electricity. As of this writing all reactors have been shutdown.

## End Notes

[1] On April 26, 1986, the worst accident in the history of civilian nuclear power occurred at the Chernobyl nuclear power plant in Ukraine, where an explosion destroyed the core of a reactor containing approximately 200 tons of nuclear fuel. The explosion also destroyed much of the reactor building, severed the reactor's cooling pipes, and spewed hot fragments of reactor fuel from the core. The explosion dispersed radioactive material over 60,000 square miles of land primarily in Ukraine, Belarus, and Russia. A separate fund was established to help stabilize the damaged reactor at Chernobyl by constructing a new containment structure. We reported in 2007 that the estimated cost of this effort was $1.2 billion, of which the United States pledged $203 million. See GAO, *Nuclear Safety: Construction of the Protective Shelter for the Chernobyl Nuclear Reactor Faces Schedule Delays, Potential Cost Increases, and Technical Uncertainties*, GAO-07-923 (Washington, D.C.: July 19, 2007).

[2] In 2010, Germany amended its Atomic Energy Act to align with an "energy concept" where nuclear power would serve a "bridging function" until an infrastructure for renewable fuel sources was reliably in place, without setting a timeline for phasing out nuclear power. On June 30, 2011, after the Fukushima Daiichi accident, the German parliament voted to fully shut down its nuclear power plants by the end of 2022. This vote followed the suspension of operations of 8 of Germany's 17 nuclear power plants.

[3] The National Diet of Japan—the Japanese parliament—commissioned an investigation of the Fukushima Daiichi accident, chaired by Kiyoshi Kurokawa, known formally as The National Diet of Japan Fukushima Nuclear Accident Independent Investigation Commission. The commission issued its findings on June 9, 2012.

[4] Safety culture implies individual and organizational awareness of and commitment to the importance of safety. It also refers to the personal dedication and accountability of all individuals engaged in any activity that has a bearing on the safety of nuclear power plants. We reported in 2010 that development of a positive safety culture often involves a shift in how workers view and address safety-related events. See GAO, *Biological Laboratories: Design and Implementation Considerations for Safety Reporting Systems*, GAO-10-850 (Washington, D.C.: Sept. 10, 2010).

[5] GAO, *Nuclear Commerce: Governmentwide Strategy Could Help Increase Commercial Benefits from U.S. Nuclear Cooperation Agreements with Other Countries*, GAO-11-36 (Washington: D.C.: Nov. 4, 2010).

[6] IAEA is an autonomous international organization affiliated with the United Nations, established in Vienna, Austria, in 1957. The agency has the dual role of promoting the peaceful uses of nuclear energy by transferring nuclear science and technology through its nuclear science and applications and technical cooperation programs, and verifying, through its safeguards program, that nuclear material subject to safeguards is not diverted to nuclear weapons or other proscribed purposes. The IAEA also develops (nonbinding) international standards for nuclear safety, as well as criteria for nuclear regulatory independence.

[7] When we originally accepted this request, Senator Carper was Chairman of the Subcommittee on Clean Air and Nuclear Safety, Committee on Environment and Public Works. In February 2014, Senator Carper became Chairman of the Subcommittee on Transportation and Infrastructure. As a result, we are addressing this report to him in his current capacity.

[8] The 16 countries we selected are Argentina, Armenia, Belgium, Canada, China, France, Indonesia, Japan, Pakistan, Russia, South Korea, Sweden, United Arab Emirates, United Kingdom, United States, and Vietnam.

[9] Because this is a nonprobability sample, information from these countries is not representative of the entire population of established and emerging civilian nuclear power countries and cannot be generalized. However, this information can provide illustrative examples of these countries' nuclear regulatory framework.

[10] We included countries with emerging or potential nuclear programs in our sample because the implications of the Fukushima Daiichi accident extend to such countries.

[11] These meetings include the April 2011 Convention on Nuclear Safety review meeting; the August 2012 Convention Second Extraordinary Meeting; and the International Atomic Energy Agency's April 2013 International Conference on Effective Nuclear Regulatory Systems.

[12] The OECD is an organization of 34 countries that work together to address the economic, social, and environmental challenges of globalization. The organization provides a setting where governments can, among other things, compare policy experiences, seek answers to common problems, and seek to coordinate domestic and international policies.

[13] NEA is a specialized agency within OECD, an intergovernmental organization of industrialized countries based in Paris, France. In September 2013, OECD's NEA issued a report on the actions taken and lessons learned by NEA member countries and standing technical committees in response to the Fukushima Daiichi accident.

[14] After the Chernobyl disaster, nuclear power plant operators worldwide were determined to work together to ensure such a disaster could never happen again. From this, WANO emerged and formally came into being on May 15, 1989, during an inaugural meeting in Moscow. The WANO Charter is signed by 144 power companies, committing them to work in support of the WANO mission to maximize the safety and reliability of nuclear power plants worldwide by working together to assess, benchmark, and improve performance through mutual support, exchange of information, and emulation of best practice.

[15] Reactors 1, 2, and 3 were operating at the time of the accident. Reactors 4, 5, and 6 were off-line for scheduled outages.

[16] The tsunami caused by the earthquake flooded and totally destroyed all but one of the emergency diesel generators, the seawater cooling pumps, the electric wiring system, and the direct current power supply for reactors 1, 2, and 4, resulting in loss of all power except for an external supply to reactor 6 from an air-cooled emergency diesel generator. In short, reactors 1, 2, 4, and 5 lost all power. Reactor 3 first partially lost power and later lost all power.

[17] "Containment" refers to the structure and associated systems that enclose the nuclear reactor and are the final barrier to the radioactive materials that may otherwise be released into the environment in the case of an accident. Such enclosures are usually dome-shaped and made of steel-reinforced concrete.

[18] EURATOM acts in several areas connected with atomic energy, including research, safety standards, and the peaceful uses of nuclear energy. One of the fundamental objectives of the EURATOM Treaty that established EURATOM is to ensure that all users in the EU enjoy a regular and equitable supply of ores and nuclear fuels.

[19] These assessments examined how nuclear installations can withstand the consequences of various extreme external events and, in certain cases, security threats and incidents due to malevolent or terrorist acts. The topics of the tests included natural initiating events (earthquake, tsunami, and extreme climatic conditions), loss of the facility's safety systems, and severe accident management.

[20] The Doel power plant is located along the Scheldt River and has four reactors with a combined output of 2,912 megawatts.

[21] NRC defines emergency planning zones as areas for which planning is recommended to assure that prompt and effective actions can be taken to protect the public in the event of a radiological incident at a nuclear power plant.

[22] A corrective action program is an operator's process for tracking, evaluating, and resolving deficiencies.

[23] On February 15, 2013, a meteor exploded over Chelyabinsk in Russia. The Mayak nuclear facility, which is near the explosion, was not affected.

[24] "Updated Measures to Mitigate Consequences of Beyond-Design Basis Accidents at Nuclear Power Plants."

[25] In the United States, an August 23, 2011, magnitude-5.8 earthquake in Mineral, Virginia, which had its epicenter approximately 11 miles from the North Anna Power Station was a beyond design-basis event. The ground motion at the site during the earthquake exceeded levels for which the plant was originally designed and licensed. Nevertheless, the plant did not experience significant damage to its structures or safety systems.

[26] This type of hydrogen control system refers to catalytic recombiners, which oxidize the hydrogen (or "recombine" it with oxygen to make water or steam).

[27] There are three major types of containment designs for boiling water reactors: Mark I, Mark II, and Mark III. The Fukushima Daiichi reactor Units 1-5 had Mark I containments, and reactor Unit 6 had a Mark II containment. According to a commissioner vote, Mark I and II containment structures' relatively small volume presents "long-recognized vulnerabilities" in their ability to contain radioactive materials during a severe accident. A 1975 NRC study (NUREG-75/014) found that the risk of containment failure during severe accidents is higher at boiling water reactors with Mark I containments because the containment volume of Mark I containment designs was significantly less than that of the other containment designs, approximately one-sixth the volume of large dry pressurized water reactor containments. In boiling water reactors with a Mark II containment design, the containment volume could be approximately 25 percent larger than the volume of Mark I containments. According to NRC's Near-Term Task Force, it can reasonably be concluded that under circumstances similar to those of the Fukushima-Daiichi accident, a Mark II would have suffered similar consequences because its containment design is only slightly larger in volume than Mark I containment designs.

[28] Generally, federal regulations, including those of NRC, can only be changed through the rulemaking process. An order—which is issued immediately—bypasses NRC's rulemaking process, which could take years and seeks comments from stakeholders, such as industry and the public. NRC's "backfit rule" requires, in order to impose new requirements on existing licensees, that NRC determine that the new requirements would result in a substantial increase in the overall protection of public health and safety or common defense and security and that this increased protection justifies the cost of implementation, unless the new requirement fits within one of the exceptions to the backfit rule in 10 C.F.R. § 50.109(a)(4)(i)-(iii). Backfitting is defined by NRC as the modification of or addition to systems, structures, components, or design of a facility; or the design approval or manufacturing license for a facility; or the procedures or organization required to design, construct, or operate a facility; any of which may result from a new or amended provision in NRC's regulations or the imposition of a regulatory staff position interpreting NRC's regulations that is either new or different from a previously applicable staff position, relative to specific dates, such as issuance of licenses.

[29] According to Swedish nuclear regulatory officials, Sweden's nuclear regulatory body's nuclear safety work is funded primarily by user fees. The increased allocation since the Fukushima Daiichi accident reflects an increase in the amount of the budget funded by the government.

[30] Even before the 2006 law that established ASN as an independent commission, the French agencies responsible for nuclear regulation became more independent over time. The first nuclear regulatory body was within the French Atomic Energy Commission, which also operated the reactors.

[31] A Severe Accident Management Program is a program to prevent and mitigate the consequences of beyond design-basis accidents, including severe accidents. According to IAEA, accident management means taking actions during a beyond design-basis accident to prevent it from escalating into a severe accident, mitigate any consequences, and achieve a long-term safe stable state.

[32] NSSC found 215 cases of quality-record falsification among 20 suppliers of reactor components for the Yeonggwang nuclear power plant. Another incident of concern to NSSC was a safety failure at the Kori nuclear power plant that was not reported to NSSC until a month after the incident. According to Korean nuclear regulatory officials, both incidents had an impact on public perceptions of nuclear safety.

[33] Under the quasi-autonomous nongovernmental organization construct, an agency can perform functions of the government and be entirely funded by the government but not controlled directly by the central government.

[34] Normally, alternating current power, which is essential for safe operation and accident recovery at commercial (civilian) nuclear power plants, is supplied by off-site sources via the electrical grid. Loss of this off-site power can have a major negative impact on a power plant's ability to achieve and maintain safe shutdown conditions.

[35] A VPN is a private network that is maintained across a shared or public network, such as the Internet, by means of specialized security procedures. VPNs are intended to provide secure connections between remote clients, such as branch offices or traveling personnel and a central office.

[36] The Nuclear Safety Directive of the European Commission requires all EU member states to host an IRRS mission and a follow-up mission once every 10 years.

[37] IAEA has conducted IRRS missions since 1992. The present form of IRRS mission dates to 2006.

[38] In the case of IRRS missions, and in accordance with IRRS guidelines, countries are encouraged to make their reports available to the public. IAEA will make a summary report publicly available 90 days after the letter transmitting the report to the host country unless the country specifically requests that it remain restricted.

[39] The Kurokawa Commission scrutinized the relationship between the plant operators and the regulators and found the relationship lacking in independence, transparency, and safety culture, calling the relationship a "typical example of regulatory capture in which the oversight of the industry by regulators effectively ceases."

[40] In addition to the 14 EU member states with operating nuclear power plants (Belgium, Bulgaria, Czech Republic, Finland, France, Germany, Hungary, Netherlands, Romania, Slovakia, Slovenia, Spain, Sweden, United Kingdom), and two neighboring states (Switzerland and Ukraine), Lithuania also participated in the comprehensive safety risk assessments even though the two reactors at its Ignalina nuclear power plant are currently being decommissioned.

[41] An ultimate heat sink is a source of water near the nuclear plant—a lake, river, or ocean, or a dedicated water source— to absorb decay heat. An ultimate heat sink is required to provide

all of the nuclear power reactor's cooling water needs for the first 30 days of an accident. Two or more pumps are used to take water from the heat sink to the plant.

[42] ENSREG is an independent, authoritative expert body composed of senior officials from national regulatory or nuclear safety authorities from all member states in the EU. WENRA is an organization composed of the chief nuclear regulators of EU countries with nuclear power plants and other interested European countries. WENRA's main objectives are to facilitate the exchange of nuclear safety information and experience among regulators, develop a common approach to nuclear safety, and provide an independent capability to examine nuclear safety in affiliated countries.

[43] According to the Convention's national report guidelines, their purpose is to suggest that material that may be useful for parties to include in their reports and to facilitate the most efficient review of implementation by the parties of their obligations under the Convention.

[44] GAO, *Nuclear Safety: Convention on Nuclear Safety Is Viewed by Most Member Countries as Strengthening Safety Worldwide*, GAO-10-489 (Washington, D.C.: Apr. 29, 2010).

[45] Of the 75 parties to the Convention, and that were obligated to submit a national report in 2012, 61 prepared such reports.

## End Notes for Appendix I

[1] Regulatory frameworks may differ in that a nuclear regulatory agency could be a cabinet agency or ministry, or an independent commission. We also considered structural changes undertaken by nuclear regulatory agencies from one type to another (e.g., agency to commission).

[2] This figure includes reactors that are temporarily suspended and thus not "operating," including the 50 reactors in Japan suspended in the aftermath of the Fukushima Daiichi accident. The three "emerging" categories are in varying stages of considering, planning, or licensing their nuclear power programs; for example, the Indonesian government has not yet decided to pursue a nuclear program but had stated an interest in developing one.

[3] These meetings include the April 2011 Convention on Nuclear Safety review meeting; the August 2012 Convention Second Extraordinary Meeting; and the International Atomic Energy Agency's April 2013 International Conference on Effective Nuclear Regulatory Systems.

In: Nuclear Safety ...
Editor: Glenn Freeman

ISBN: 978-1-63482-280-0


*Chapter 2*

# FUKUSHIMA NUCLEAR DISASTER*

***Mark Holt, Richard J. Campbell and Mary Beth Nikitin***

## SUMMARY

The huge earthquake and tsunami that struck Japan's Fukushima Daiichi nuclear power station on March 11, 2011, knocked out backup power systems that were needed to cool the reactors at the plant, causing three of them to undergo fuel melting, hydrogen explosions, and radioactive releases. Radioactive contamination from the Fukushima plant forced the evacuation of communities up to 25 miles away and affected up to 100,000 residents, although it did not cause any immediate deaths.

Tokyo Electric Power Company (TEPCO) operates the Fukushima nuclear power complex in the Futaba district of Fukushima prefecture in Northern Japan, consisting of six nuclear units at the Fukushima Daiichi station and four nuclear units at the Fukushima Daini station. All the units at the Fukushima complex are boiling water reactors, with reactors 1 to 5 at the Fukushima Daiichi site being the General Electric Mark I design, which is also used in the United States. The Fukushima Daiichi reactors entered commercial operation in the years from 1971 (reactor 1) to 1979

---

* This is an edited, reformatted and augmented version of a Congressional Research Service publication R41694, prepared for Members and Committees of Congress dated January 18, 2012.

(reactor 6). The Fukushima Daini reactors shut down automatically after the earthquake and were able to maintain sufficient cooling.

When the earthquake struck, Fukushima Daiichi units 1, 2, and 3 were generating electricity and shut down automatically. The earthquake caused offsite power supplies to be lost, and backup diesel generators started up as designed to supply backup power. However, the subsequent tsunami flooded the electrical switchgear for the diesel generators, causing most AC power in units 1 to 4 to be lost. Because Unit 4 was undergoing a maintenance shutdown, all of its nuclear fuel had been removed and placed in the unit's spent fuel storage pool. One generator continued operating to cool units 5 and 6.

The loss of all AC power in units 1 to 3 prevented valves and pumps from operating that were needed to remove heat and pressure that was being generated by the radioactive decay of the nuclear fuel in the reactor cores. As the fuel rods in the reactor cores overheated, they reacted with steam to produce large amounts of hydrogen, which escaped into the unit 1, 3, and 4 reactor buildings and exploded (the hydrogen that exploded in Unit 4 is believed to have come from Unit 3). The explosions interfered with efforts by plant workers to restore cooling and helped spread radioactivity. Cooling was also lost in the reactors' spent fuel pools, although recent analysis has found that no significant overheating took place.

Radioactive material released into the atmosphere produced extremely high radiation dose rates near the plant and left large areas of land uninhabitable, especially to the northwest of the plant. Contaminated water from the plant was discharged into the sea, creating international controversy.

The United States and other countries, as well as the International Atomic Energy Agency, are providing assistance to Japan to deal with the nuclear disater. U.S. assistance has included transport of pumps, boron, fresh water, remote cameras, use of Global Hawk surveillance drones, evacuation support, medical support, and decontamination and radiation monitoring equipment.

Studies of the Fukushima disaster have identified design changes, response actions, and other safety improvements that could have reduced or eliminated the amount of radioactivity released from the plant. As a result, Fukushima has prompted a reexamination of nuclear plant safety requirements around the world, including in the United States.

## INTRODUCTION AND OVERVIEW

The disaster that struck Japan's Fukushima Daiichi nuclear power station on March 11, 2011, caused the most extensive release of radioactivity since the Chernobyl accident in 1986 and was far worse than the 1979 Three Mile Island accident in the United States. Unlike at Chernobyl and Three Mile Island, the destruction at Fukushima was initiated by natural disasters—a huge earthquake and tsunami—rather than equipment failure and human error. The tsunami knocked out backup power systems that were needed to cool the reactors at the plant, causing several of them to undergo fuel melting, hydrogen explosions, and radioactive releases.

Studies of the Fukushima disaster have identified design changes, response actions, and other safety improvements that could have reduced or eliminated the amount of radioactivity released from the plant. As a result, Fukushima has prompted a reexamination of nuclear plant safety requirements around the world, including the United States.

Radioactive contamination from the Fukushima plant forced the evacuation of communities up to 25 miles away, affecting up to 100,000 residents, many of whom remain indefinitely barred from their homes. The evacuations are believed to have prevented radiation exposure among the population from exceeding Japanese regulatory limits in most cases. Near-term deaths and illnesses resulting from radiation are believed to be unlikely, although cancer and other long-term health effects remain possible. Workers at the plant site were exposed to far higher radiation levels, with at least two suffering radiation burns on their legs after wading in contaminated water. Two other workers drowned during the tsunami.

Recovery from the disaster has focused on restoring cooling systems at the three most severely damaged reactors at the six-unit plant and halting radioactive emissions into the air and water. That work has been hampered by continuing high levels of radiation in the plant and severe structural damage. The Japanese government announced December 16, 2011, that the damaged Fukushima reactors had achieved "cold shutdown," a milestone in which the reactor cooling water is below boiling temperature at atmospheric pressure. At cold shutdown, the diminished threat of further radioactive releases may allow some residents to begin returning to the least contaminated evacuation zones.[1]

Japan's environment minister announced December 19, 2011, that about $15 billion had been allocated for decontaminating the area around the Fukushima Daiichi plant, an unprecedented undertaking.[2] Complete decommissioning and dismantlement of the plant is expected to take 40 years, and the total cost of the disaster was recently estimated by a Japanese government committee to exceed $75 billion.[3]

The Institute of Nuclear Power Operations (INPO), a safety organization established by the U.S. nuclear power industry after the Three Mile Island accident, issued a detailed description of the Fukushima accident in November 2011. The INPO report provides timelines of the response actions taken at each unit of the Fukushima Daiichi plant and the sequences of events that led to major reactor core damage and radioactive releases. It is intended "to provide an accurate, consolidated source of information" about the event. However, the report notes, "Because of the extensive damage at the site, some details of the event remain unknown or have not been confirmed."[4]

The purpose of this CRS report is to highlight the aspects of the Fukushima disaster that could bear on U.S. nuclear plant safety and nuclear energy policy in general. It provides a brief explanation of the Fukushima events, including new details provided by the INPO report, a general discussion of the consequences of the disaster, and a description of U.S. assistance provided to Japan.

## SUMMARY OF THE FUKUSHIMA DAIICHI DISASTER

The earthquake on March 11, 2011, off the east coast of Honshu, Japan's largest island, reportedly caused an automatic shutdown of 11 of Japan's 55 operating nuclear power plants.[5] Most of the shutdowns proceeded without incident. However, the plants closest to the epicenter, Fukushima and Onagawa (see *Figure 1*), were damaged by the earthquake and resulting tsunami. The Fukushima Daiichi plant subsequently suffered hydrogen explosions and severe nuclear fuel damage, releasing significant amounts of radioactive material into the environment.

Tokyo Electric Power Company (TEPCO) operates the Fukushima nuclear power complex in the Futaba district of Fukushima prefecture in Northern Japan, consisting of six nuclear units at the Fukushima Daiichi station and four nuclear units at the Fukushima Daini station. All the units at the Fukushima complex are boiling water reactors (BWRs),[6] with reactors 1 to 5 at the Fukushima Daiichi site being the General Electric Mark I design (see *Figure 2*).[7] The Fukushima Daiichi reactors entered commercial operation in the years from 1971 (reactor 1) to 1979 (reactor 6).

Nuclear reactors produce power through the fission (splitting) of the nuclei of heavy isotopes, such as uranium-235 and plutonium-239, resulting from the absorption of neutrons. Each fission event generates additional neutrons that induce more fission events, creating a continuous nuclear chain reaction.

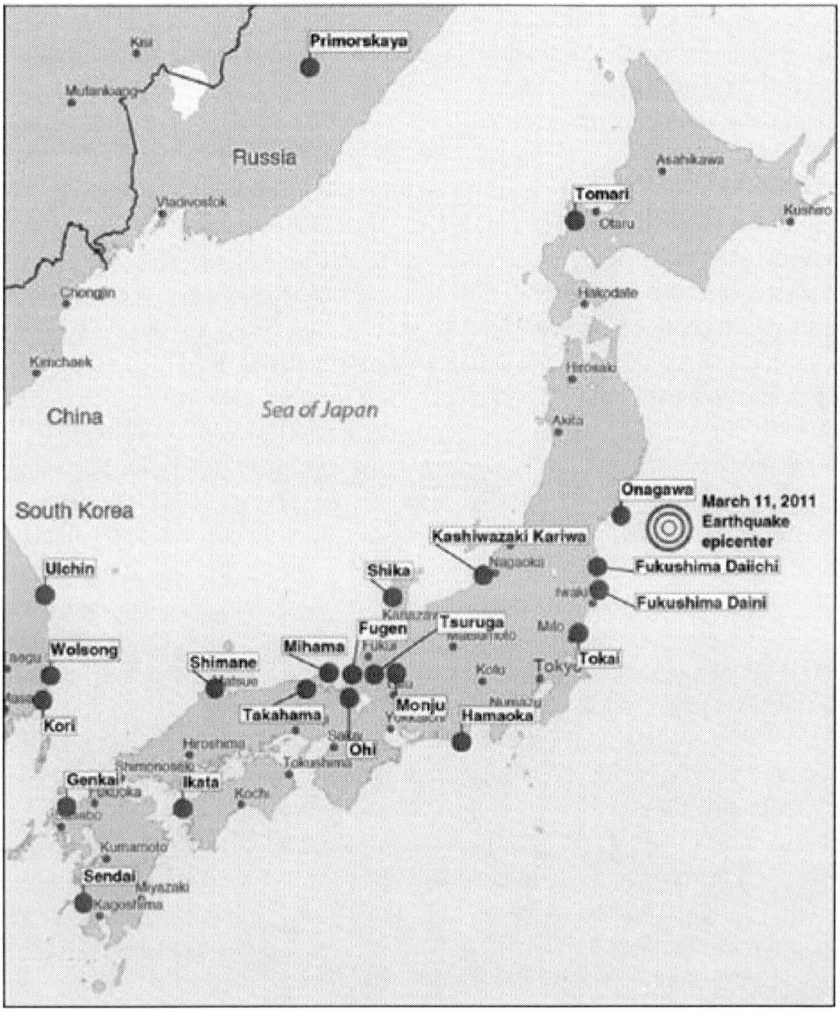


Source: Nuclear Energy Institute, edited by CRS.
Notes: http://i1107.photobucket.com/albums/h384/reactor1/japan_map1.jpg.

Figure 1. Japan Earthquake Epicenter and Nuclear Plant Locations.

The heavy nuclei split into lighter isotopes called fission products, many of which are highly radioactive, such as iodine-129, iodine-131, strontium-90, and cesium-137. To shut down the nuclear chain reaction, neutron-absorbing control rods[8] are inserted into the reactor core. However, even though the fission process has stopped, the fission products and other radioactive isotopes in the reactor core continue to generate significant heat through radioactive decay. Until the decay heat sufficiently diminishes, a source of electricity is needed to operate pumps and circulate water in the reactor. Under normal conditions, it would take a few days for a reactor core to cool down to a "cold shutdown" state.[9]

Reactors 1, 2, and 3 at Fukushima Daiichi were operating and automatically shut down when the quake struck, while reactors 4, 5, and 6 were already shut down for routine inspections. All four of the Fukushima Daini reactors were operating at the time of the earthquake and taken down after the quake. Although horizontal ground acceleration from the magnitude 9.0 earthquake exceeded the maximum designed level at two of the operating Fukushima Daiichi units (2 and 3), all the operating units maintained normal cooling immediately after the emergency shutdown.[10]

However, the earthquake triggered a tsunami that struck the Fukushima Daiichi station about 40 minutes later, devastating much of the area and overtopping the plant's six-meter-high seawall. TEPCO estimated the tsunami's height at Fukushima Daiichi to be 14 meters (46 feet).[11] The station was cut off from Japan's national electricity grid, leaving the plant dependent on backup diesel generators for alternating current (AC) power.[12] The tsunami flooded most of the generators and electrical switchgear rooms, knocking out the backup AC power for cooling the nuclear reactors in units 1-5. An air-cooled diesel generator at Unit 6 survived the tsunami and provided backup power to cool unit 5 and 6, which therefore did not suffer fuel damage.[13]

Although backup batteries at the Fukushima Daiichi plant were designed to provide direct current (DC) power for eight hours, most of the DC power was lost when the tsunami flooded the plant's DC power distribution systems. As a result, the shared control room for units 1 and 2 went completely dark, and the control room for units 3 and 4 had only emergency lighting available.[14]

TEPCO immediately began to experience problems cooling the Fukushima Daiichi reactors. With alternating current no longer available to power the primary and secondary cooling systems, and batteries for backup cooling and control systems out of service, TEPCO began trying to cool the reactor cores with seawater.[15] Neutron-absorbing boron[16] was added to the seawater to prevent restart of the nuclear chain reaction. Despite those efforts, cooling water levels in the reactor cores remained low for many days, causing the reactors to overheat.

TEPCO estimated on December 2, 2011, that all of the nuclear fuel in Unit 1 had melted and that much of it had leaked from the reactor pressure vessel into the primary containment.

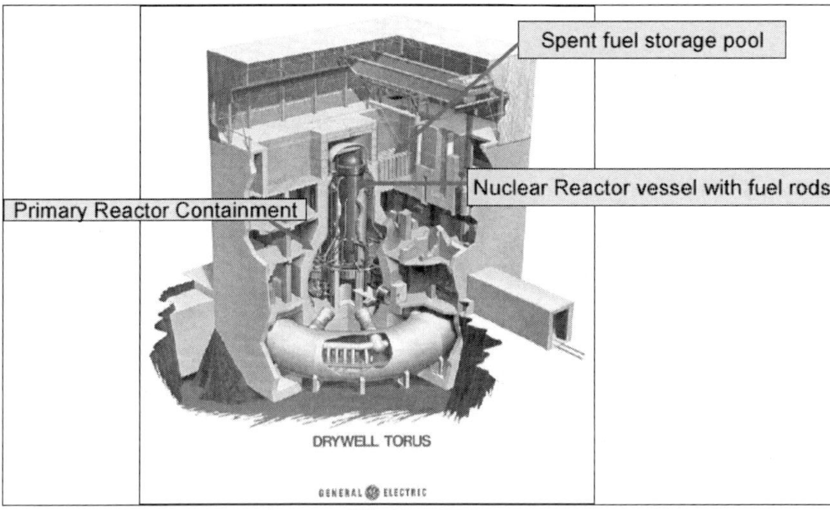


Source: http://www.nrc.gov/, as modified by CRS.

Figure 2. General Electric Mark I Boiling Water Reactor and Containment Building.

Melting was considered likely for 57% of the fuel in Unit 2 and 63% of Unit 3. Because heat removal systems failed, pressure built up to high levels in the primary containments of units 1 through 3. The containments were vented to the atmosphere to relieve the pressure, allowing the escape of radioactivity leaking from the reactor cores.[17]

A chemical reaction between the fuel's zirconium cladding and high-temperature steam is believed to have generated large amounts of hydrogen in the containments of units 1 through 3. The hydrogen leaked from the containments or the venting systems into the reactor buildings and caused large explosions in units 1, 3, and 4. Japanese safety regulators estimated that each unit had produced 800-1,000 kilograms of hydrogen.[18] The hydrogen explosions caused tremendous damage and hindered TEPCO's efforts to restore plant cooling capability.

The loss of cooling also affected the plant's spent fuel pools (shown in *Figure 2*), which hold fuel rods that have been removed from the reactors after their ability to sustain a nuclear chain reaction has diminished. Although much of the radioactivity in the spent fuel has been decaying for many years, the large volumes of spent fuel in the pools represent a significant total heat load. If water in the spent fuel pools boils away or leaks out, the spent fuel rods may overheat and release radioactive material into the air. Because of concern that water in the Fukushima Daiichi spent fuel pools may have been splashed out or leaked during the earthquake and begun boiling away after cooling systems were lost, dramatic efforts were made to spray water into the pools or drop water from helicopters. Later analysis indicated that the pools did retain sufficient cooling water during the accident.[19] However, the explosions of reactor buildings 1, 3, and 4 exposed those pools to the atmosphere, and debris from the explosions may have fallen into the pools and damaged the stored fuel.

Substantial releases of radioactive material have occurred at the plant, most likely from leaking or venting from the primary containment structure that surrounds the reactor pressure vessel. Radioactive contamination exceeding regulatory limits was found in seawater around the plant, as well as contamination of agricultural products exceeding legal standards in surrounding prefectures.[20] Radioactive contamination in Tokyo drinking water on March 23 was measured at "more than twice the accepted level for infants."[21] As discussed later in this report, depositions of radioactive cesium "revealed high values comparable with the most contaminated areas of Chernobyl, even beyond the initial 20 km-radius evacuation zone around the Fukushima plant."[22]

## Status of the Fukushima-Daiichi Reactors

All units of the plant were reconnected to off-site electrical power by March 23, and plant equipment was gradually reactivated. Diesel-generated backup power had been available at units 5 and 6 since March 19. Top priorities were restoring core cooling to units 1-3 and to the spent fuel pools in units 1-4 and eliminating discharges of highly contaminated water into the ocean.[23] All the reactors were declared to be in "cold shutdown" as of December 16, 2011.

### *Unit 1*

Unit 1 was generating electricity when the earthquake occurred and shut down automatically, but the resulting tsunami halted emergency core cooling. Water levels in the reactor vessel dropped below the top of the hot fuel, and steam began reacting with the zirconium fuel cladding to produce large amounts of hydrogen. After about 12 hours, pressure in the primary containment had increased to twice the designed level, radiation in the plant rose sharply, and preparations were made to vent the containment to the atmosphere. The population within 10 kilometers of the plant was evacuated, and venting began almost 24 hours into the emergency. About an hour after venting began on March 12, a large hydrogen explosion occurred in the upper area of the reactor building, causing widespread damage. The hydrogen may have leaked from the primary containment into the upper building structure during the overpressurization or from the vents.[24]

Plant workers began injecting seawater into the reactor pressure vessel on March 12 through a fire extinguisher line, and off-site power was restored on March 20. Freshwater injection into the reactor vessel began March 29.

TEPCO estimated on November 30, 2011, that all of the nuclear fuel in Unit 1 had melted and fallen to the bottom of the reactor pressure vessel (RPV) and that "a significant amount of the fuel fell out into the primary containment vessel (PCV) by destroying the RPV." At the same time, TEPCO estimated that molten fuel may have eaten into the concrete at the bottom of the primary containment to a depth of up to 70 centimeters, or about 28 inches.[25]

### *Unit 2*

Unit 2 was generating electricity and automatically shut down during the earthquake. The tsunami flooded one backup diesel generator, but a second, air-cooled generator was on higher ground and continued to operate. However, the electrical switchroom for the operating generator was located below ground, suffered water damage, and subsequently failed. A portable generator was brought to Unit 2 and temporary power cables were hooked up, but just before power could be restored, the hydrogen explosion in Unit 1 propelled debris into the portable generator and cables and rendered them inoperable.

On March 15, while workers were attempting to vent the Unit 2 containment, they heard a loud noise near the torus (see *Figure 2*) followed by a reduction in containment pressure. The sound was later determined not to have been an explosion as initially suspected, but it may have been associated with a breach of the primary containment. Meanwhile, the rupture of a large, rectangular blowout panel in the outer wall of Unit 2 apparently allowed hydrogen to escape from the reactor building, preventing the explosions that occurred in the adjacent units.[26]

A steam-driven backup cooling system in Unit 2 continued to operate for about 70 hours after the tsunami, but all core cooling was subsequently lost for more than six hours, allowing the hot reactor core to become uncovered by water, possibly completely, and begin to melt. Seawater injection into the reactor vessel began March 14, after being delayed by a hydrogen explosion in Unit 3. Offsite power was restored on March 20, and freshwater injection into the reactor vessel began March 26.[27] TEPCO estimates that 57% of the fuel in Unit 2 melted during the period when cooling was lost, and some of the melted fuel may have dropped to the bottom of the reactor pressure vessel. However, as of November 30, TEPCO maintained that most of the melted fuel had remained in the reactor vessel and had not fallen into the primary containment vessel as in Unit 1.[28]

### *Unit 3*

Unit 3 was generating electricity and shut down automatically during the earthquake and then lost AC power during the tsunami. High-pressure injection of water into the reactor vessel was lost after about 36 hours, on March 13, and there was no cooling for nearly the next seven hours, when seawater injection began. During that seven-hour period, the reactor core became uncovered with water, resulting in severe fuel damage and the generation of large amounts of hydrogen. Damage to the reactor pressure vessel is considered likely.[29]

Pressure in the primary containment structure rose sharply on March 14, and after many hours, plant workers opened the necessary valves to vent the primary containment at about 8:30 a.m. A large hydrogen explosion occurred at about 11 a.m. that destroyed the secondary containment (the upper part of the reactor building), where hydrogen had apparently escaped from the primary containment or backflowed from the earlier venting. The explosion injured 11 workers and damaged portable generators, fire engines, hoses, and temporary power cables that had been brought in to restore cooling to the reactor. An undamaged fire engine was brought in and seawater injection resumed at about 4:30 p.m.[30]

Unit 3 had operated since September 2010 with plutonium-based fuel,[31] called mixed-oxide (MOX) fuel. The use of MOX fuel, which made up about 6% of the reactor core, heightened concerns after the accident about the potential release of plutonium. Although plutonium, a hazardous radioactive element, is also created during irradiation of conventional nuclear fuel, there is substantially more in MOX fuel than in conventional fuel. The Nuclear Energy Institute contends that the relatively small amount of MOX fuel in Unit 3 "did not have a significant impact on the offsite releases of radioactivity."[32]

Three workers received high doses of radiation from contaminated water while installing cables in the basement of the turbine building March 24. Injection of fresh water into the reactor vessel began March 25. TEPCO estimates that, as in Unit 2, much of the reactor core of Unit 3 melted but was retained in the reactor vessel.[33] The reactor's primary containment structure is not believed to be damaged, and damage has not been found to the spent fuel in the spent fuel pool.

### *Unit 4*

Unit 4 was out of service for maintenance when the earthquake struck. All its nuclear fuel had been moved to the spent fuel pool, which eliminated the need for cooling the reactor core but greatly increased the spent fuel pool's heat load. A hydrogen explosion severely damaged the reactor building on March 15. Because there had been no fuel in the reactor core to generate the hydrogen, suspicion initially focused on the spent fuel pool as the source. Early reports had indicated that most or all of the water in the Unit 4 spent fuel pool had been lost, potentially causing overheating that could lead to hydrogen generation. However, it was later determined that water in the pool had not dropped below the level of the spent fuel and that no significant overheating had occurred. Spraying of water into the spent fuel pool began on March 20.

Analysis by TEPCO now indicates that the hydrogen that exploded in Unit 4 came from Unit 3 during the venting that had begun the previous day. Units 3 and 4 share a single exhaust stack, and the loss of power had left Unit 4's exhaust gas valves in the open position. As a result, when the Unit 3 primary containment was vented, hydrogen and other gases were able to flow backward from the common exhaust stack into Unit 4 and eventually into the reactor building's air duct system. This scenario was supported by radiation samples from the Unit 4 exhaust gas filters that showed higher contamination levels in the filters closest to the common stack (and closest to Unit 3), the opposite of what would normally be expected.[34]

### *Units 5 and 6*

Units 5 and 6, which are located separately from units 1-4, were not operating during the earthquake. Four of the five emergency diesel generators at units 5 and 6 were rendered inoperable by the tsunami, but one air-cooled diesel generator continued working. The working generator supplied continuous power to Unit 6 and later provided power to Unit 5 as well.[35] Cold shutdown of both units was declared on March 20. Holes were opened in the roofs of the reactor buildings to prevent hydrogen buildup. No other damage has been reported to the reactor buildings or spent fuel.

### Fukushima Daini

The Fukushima Daini station is approximately 12 kilometers (7 miles) south of the Daiichi station, and further removed from the epicenter of the earthquake. The earthquake and tsunami apparently caused damage to the emergency core cooling systems at reactors 1, 2, and 4, while reactor 3 was apparently able to shut down without problems. The station reportedly retained offsite power to maintain its ability to circulate cooling water in the reactor. The makeup water and condensate systems were used as an emergency measure to maintain cooling water levels in reactors 1, 2, and 4. TEPCO has since made repairs to the cooling systems, and stable, cold shutdown conditions were reported at all Daini reactors on March 14, 2011.[36]

## RADIOACTIVE RELEASES AND CONTAMINATION

Radioactive material escaping from the damaged Fukushima-Daiichi reactor cores created high radiation levels throughout the plant and eventually caused widespread contamination. Large radioactive releases into the atmosphere resulted from the containment venting and hydrogen explosions. At the same time, large amounts of water being pumped or sprayed into the reactors became heavily contaminated and were discharged or leaked into the ocean.

A precautionary evacuation for a 3-kilometer radius around the plant was ordered on the day of the tsunami. Elevated dose rates were first detected at the plant site boundary on the next day, and the evacuation was expanded first to 10 kilometers and then to 20 kilometers. As the situation in the reactors deteriorated, dose rates at the site boundary rose rapidly, reaching 1,085 millirem/hour on March 16. Meanwhile, dose rates inside the plant were measured at up to 400 times higher, greatly complicating recovery efforts.[37] For comparison, a widely used international dose limit for members of the public is 100 millirem for an entire year.

Radioactive contamination of the ocean became an international concern after high concentrations of radioactivity were measured in the power plant's harbor on April 2. According to the INPO report, contaminated water had accumulated in the turbine building, flowed through a trench, and leaked into the harbor, totaling about 130,000 curies.[38]

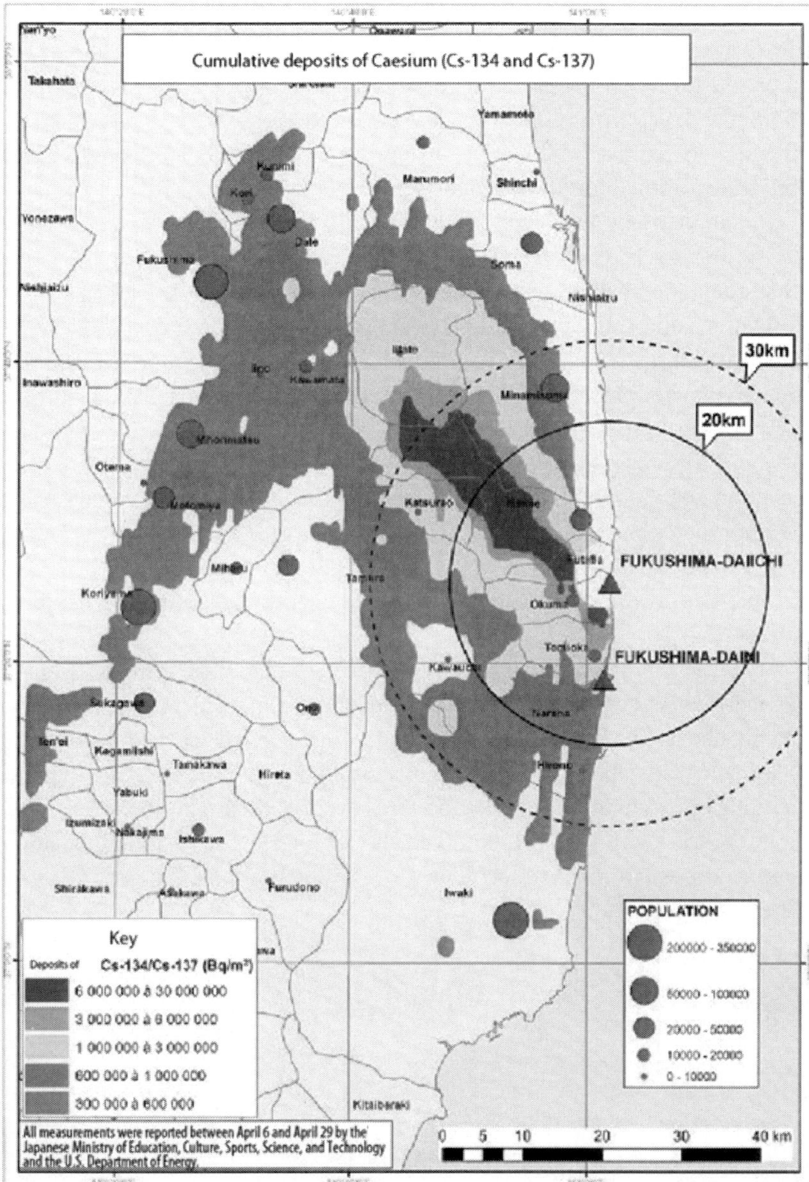


Source: Institut de Radioprotection et de Surete Nucleaire (IRSN)
Note: 1 becquerel (Bq) = 27 picocuries (pCi)

Figure 3. Cesium Depositions Around Fukushima-Daiichi.

Controversy was raised by Japan's decision in early April 2011 to allow TEPCO to release more than 10,000 cubic meters of water with relatively low concentrations of radioactivity into the sea to make space to store more heavily contaminated water from the plant. Treatment facilities were since constructed at the site to remove radioactive material and other contaminants from the stored water, allowing it to be recycled for reactor cooling.[39]

Contamination of the land around the Fukushima plant was concentrated to the northwest, leaving large areas uninhabitable, as shown in *Figure 3.* Most of that contamination occurred when prevailing winds that had been carrying the plant's atmospheric releases over the Pacific Ocean shifted to the northwest during two days in mid-March and heavy rains washed the radioactive material from the air.[40] The depositions of radioactive cesium (the primary long-term land contaminant) are "comparable with the most contaminated areas of Chernobyl, even beyond the initial 20 km-radius evacuation zone around the Fukushima plant," according to the French Institut de Radioprotection et de Surete Nucleaire (IRSN). However, the amount of land around Fukushima with cesium deposits greater than 600,000 becquerels (16.2 microcuries) per square meter was estimated to be "only 8.5% of that of Chernobyl."[41]

About 84,000 residents within 20 kilometers (about 12.5 miles) from the plant were evacuated by April, and another 15,000 residents in the heavily contaminated area northwest of the 20- kilometer zone, up to 40 kilometers (25 miles) from the plant, were evacuated in mid-May. The Japanese government has announced plans to begin decontaminating the area, but how long the process will take is unknown. The long-term goal is to reduce annual radiation doses from the accident to the international standard of 100 millirem.[42]

## U.S. Assistance

The United States and other countries, as well as the International Atomic Energy Agency, are providing assistance to Japan to deal with the nuclear crisis. Japan requested assistance for the nuclear crisis from the United States a few days after the earthquake and tsunami. The United States provided advice and equipment, including transport of pumps, boron, fresh water, remote cameras, use of Global Hawk surveillance drones, evacuation support, medical support, and decontamination and radiation monitoring equipment. A U.S. Nuclear Regulatory Commission (NRC) advisory team traveled to Japan at the

Japanese government's request, and NRC experts were resident at the U.S. embassy in Japan throughout the crisis. The National Nuclear Security Administration (NNSA) of the Department of Energy (DOE) sent "a team of 34 experts and more than 17,000 pounds of equipment in support of efforts to manage the crisis."[43] DOE also provided the Aerial Measuring System and radiation monitoring equipment, and provided analysis from the U.S. national laboratories. The U.S. Department of Defense provided high-pressure water pumps and fire trucks in the battle to cool the reactors and contain the damage.

The United States continues to work with Japan on environmental remediation efforts and the decommissioning of the Fukushima Daiichi plant. Deputy Secretary of Energy Daniel Poneman in a speech in Tokyo said that Japan has asked the Department of Energy to share its experience of environmental cleanup at nuclear sites such as Hanford and Savannah River.[44] The Department of Energy has participated in workshops and training for Japanese officials. For example, DOE's Office of Environmental Management conducted a workshop in Japan in October 2011 on the storage and disposal of radioactive waste.[45] This cooperation is expected to last many years into the future.

## End Notes

[1] Yuji Okada, Jacob Adelman, and Stuart Biggs, "Fukushima Dismantling to Start After Cold Shutdown," Bloomberg, December 16, 2011, http://www.bloomberg.com/news/2011-12-15/tepco-said-to-expect-approval-to-startdecommissioning-fukushima-station.html.

[2] Toko Sekiguchi, "Japan Allocates Money for Decontamination," Wall Street Journal, December 19, 2011.

[3] Shota Ushio, "Japan's Nuclear Power to Remain Competitive Through 2030: Panel," Nucleonics Week, December 22, 2011.

[4] Institute of Nuclear Power Operations, Special Report on the Nuclear Accident at the Fukushima Daiichi Nuclear Power Station, INPO 11-005, November 2011, available at http://www.nei.org/resourcesandstats/documentlibrary/ safetyandsecurity/reports/special-report-on-the-nuclear-accident-at-the-fukushima-daiichi-nuclear-power-station.

[5] "Timeline: Japan Power Plant Crisis," BBC News, March 13, 2011, http://www. bbc.co.uk/ news/science-environment12722719.

[6] A common nuclear power reactor design in which water flows upward through the core, where it is heated by fission and allowed to boil in the reactor vessel. The resulting steam then drives turbines, which activate generators to produce electrical power. BWRs operate similarly to electrical plants using fossil fuel, except that the BWRs are powered by 370–800 nuclear fuel assemblies in the reactor core rather than burning coal or natural gas to create steam. U.S. Nuclear Regulatory Commission, "Boiling-Water Reactor (BWR)," http://www.nrc.gov/reading-rm/basic-ref/glossary/ boiling-water-reactor-bwr.html.

[7] Nuclear Information and Resource Service, "Fact Sheet on Fukushima Nuclear Power Plant," http://www.nirs.org/ reactorwatch/accidents/Fukushimafactsheet.pdf.

[8] A rod, plate, or tube containing a material such as hafnium, boron, etc., used to control the power of a nuclear reactor. By absorbing neutrons, a control rod prevents the neutrons from causing further fissions. U.S. Nuclear Regulatory Commission, "Control Rod," http://www.nrc.gov/reading-rm/basic-ref/glossary/control-rod.html.

[9] U.S. Nuclear Regulatory Commission, "Cold Shutdown," http://www.nrc.gov/reading-rm/basic-ref/glossary/coldshutdown.html.

[10] Institute of Nuclear Power Operations, op. cit., p. 6.

[11] "Fukushima Faced 14-Metre Tsunami," World Nuclear News, March 23, 2011, http://www.world-nuclear-news.org/RS_Fukushima_faced_14-metre_tsunami_2303113.html.

[12] "Timeline: Japan Power Plant Crisis," BBC News, March 13, 2011, http://www.bbc.co.uk/news/scienceenvironment-12722719.

[13] Institute of Nuclear Power Operations, op. cit., p. 8.

[14] Ibid.

[15] Yanmei Xie and Steven Dolley, "Damaged Reactor Design Has Weakness, But Not Cause for Crises, Say Experts," Nucleonics Week, March 17, 2011, p. 1.

[16] Boron is the main material that goes into control rods used to halt or slow fission reactions in nuclear reactors. Japan Times Online, "Seoul to Send Boron in Bid to Cool Reactors," March 16, 2011, http://search.japantimes.co.jp/cgi-bin/ nn20110317a9.html.

[17] Shoto Ushio, "Tepco Report of Core Melting Renews Questions About Severe Accidents," Nucleonics Week, December 8, 2011, p. 1.

[18] World Nuclear Association, "Fukushima Accident 2011," December 22, 2011, citing the Japanese Nuclear and Industrial Safety Agency (NISA), http://world-nuclear.org/info/fukushima_accident_inf129.html.

[19] Institute of Nuclear Power Operations, op. cit., p. 12.

[20] Japan Atomic Industrial Forum, "Status of Nuclear Power Plants in Fukushima as of 22:00 March 24," March 24, 2011, http://www.jaif.or.jp/english/.

[21] U.S. Department of State, Bureau of Legislative Affairs, "Japan Earthquake Update 19," March 23, 2011.

[22] Institut de Radioprotection et de Surete Nucleaire, Assessment on the 66th Day of Projected External Doses for Populations Living in the North-West Fallout Zone of the Fukushima Nuclear Accident, Report DRPH/2011-10, May 27, 2011, http://www.irsn.fr/EN/news/Documents/IRSN-Fukushima-Report-DRPH-23052011.pdf.

[23] Ibid.; additional status information is from: Japan Nuclear Energy Safety Organization, "The State of Fukushima Dai-ich by the Impact of Tohoku Pacific Ocean Earthquake," March 23, 2011; Japanese Nuclear and Industrial Safety Agency, Seismic Damage Information (the 70th Release), April 3, 2011, http://www.nisa.meti.go.jp/english/files/ en20110404-2-1.pdf.

[24] Institute of Nuclear Power Operations, op. cit., p. 20.

[25] Tokyo Electric Power Company (TEPCO), "The Evaluation Status of Reactor Core Damage at Fukushima Daiichi Nuclear Power Station Units 1 to 3," November 30, 2011, pp. 6 and 21, http://www.tepco.co.jp/en/nu/fukushima-np/ images/handouts_111130_04-e.pdf.

[26] Institute of Nuclear Power Operations, op. cit., p. 28.

[27] Institute of Nuclear Power Operations, op. cit., p. 27.

[28] TEPCO, op. cit., p. 7.

[29] Institute of Nuclear Power Operations, op. cit., p. 10.

[30] Institute of Nuclear Power Operations, op. cit., p. 32; World Nuclear Association, op. cit.

[31] World Nuclear Association, "Nuclear Power in Japan," February 24, 2011, http://www.world-nuclear.org/info/ inf79.html.

[32] Nuclear Energy Institute, "What Fuel Mix Was in Use at Fukushima and Was That a Factor in the Accident?," July 25, 2011, http://safetyfirst.nei.org/ask-an-expert/what-fuel-mix-was-in-use-at-fukushima-and-was-that-a-factor-in-theaccident-does-the-united-states-use-mox-fuel/.

[33] TEPCO, op. cit., p. 7.

[34] Institute of Nuclear Power Operations, op. cit., p. 33.

[35] Institute of Nuclear Power Operations, op. cit., p. 3.

[36] "All Fukushima Daini Units in Cold Shutdown," World Nuclear News, March 14, 2011, http://www.world-nuclearnews.org/IT-All_Fukushima_Daini_units_in_cold_shutdown-1503114.html.

[37] Institute of Nuclear Power Operations, op. cit., p. 38.

[38] Institute of Nuclear Power Operations, op. cit., p. 38. Curies and becquerels are units for measuring radioactive decay. One becquerel equals one radioactive decay event per second, while one curie equals 37 billion decay events per second. Rems and seiverts are the units for measuring the effect of radiation doses absorbed by the human body.

[39] World Nuclear Association, op. cit.

[40] Institute of Nuclear Power Operations, op. cit., p. 42; Nuclear Information and Resource Service, "Chronological Fact Sheet on 2011 Crisis at Fukushima Nuclear Power Plant," Update, December 7, 2011, http://www.nirs.org/ reactorwatch/accidents/Fukushima factsheet.pdf.

[41] Institut de Radioprotection et de Surete Nucleaire, op. cit., p. 4.

[42] World Nuclear Association, op. cit.

[43] Remarks of Daniel Poneman, Deputy Secretary, Department of Energy, Tokyo American Center, December 15, 2011, http://energy.gov/articles/deputy-secretary-daniel-ponemans-remarks-tokyo-american-center-prepared-delivery.

[44] Ibid.

[45] http://www.em.doe.gov/pdfs/EM%20Program%20Update%2011-09-11.pdf.

In: Nuclear Safety …
Editor: Glenn Freeman

ISBN: 978-1-63482-280-0


*Chapter 3*

# NUCLEAR ENERGY POLICY*

*Mark Holt*

## SUMMARY

Nuclear energy issues facing Congress include reactor safety and regulation, radioactive waste management, research and development priorities, federal incentives for new commercial reactors, nuclear weapons proliferation, and security against terrorist attacks.

The earthquake and resulting tsunami that severely damaged Japan's Fukushima Daiichi nuclear power plant on March 11, 2011, raised questions in Congress about the disaster's possible implications for nuclear safety regulation, U.S. nuclear energy expansion, and radioactive waste policy. The tsunami knocked out electric power at the six-reactor plant, resulting in the overheating of several reactor cores, loss of cooling in spent fuel storage pools, major hydrogen explosions, and releases of radioactive material into the environment. The Nuclear Regulatory Commission (NRC) issued orders to U.S. nuclear plants March 12, 2012, to begin implementing safety improvements in response to Fukushima.

Significant incentives for new commercial reactors were included in the Energy Policy Act of 2005 (EPACT05, P.L. 109-58), such as tax credits and loan guarantees. Together with volatile fossil fuel prices and the possibility of greenhouse gas controls, the federal incentives for nuclear power helped spur renewed interest by utilities and other potential reactor developers. License applications for as many as 31 new reactors

* This is an edited, reformatted and augmented version of a Congressional Research Service publication, No. RL33558, dated October 15, 2014.

were announced, and NRC issued licenses for four reactors at two plant sites in early 2012. However, falling natural gas prices, safety concerns raised by the Fukushima accident, and other changing circumstances have made it unlikely that many more of the proposed nuclear projects will move toward construction in the near term.

Four U.S. reactors were permanently closed in 2013, and another shutdown has been announced for late 2014. Three reactors were closed because of the need for major repairs, and the other two because electricity prices fell below their generating costs.

DOE's nuclear energy research and development program includes advanced reactors, fuel cycle technology and facilities, and infrastructure support. The Obama Administration's FY2015 funding request for nuclear energy research and development totaled $863.4 million. Including advanced reactors, fuel cycle technology, infrastructure support, and safeguards and security, the total nuclear energy request was $25.0 million (2.9%) below the FY2014 funding level. The House approved $826.7 million (H.R. 4923), while the Senate Subcommittee on Energy and Water Development Appropriations recommended $777.0 million.

Disposal of highly radioactive waste has been one of the most controversial aspects of nuclear power. The Obama Administration halted work on a long-planned waste repository at Yucca Mountain, NV, and established the Blue Ribbon Commission on America's Nuclear Future (BRC) to recommend new approaches to the waste problem. The BRC issued its final report to the Secretary of Energy on January 26, 2012. In response to the BRC report, and to provide an outline for a new nuclear waste program, DOE issued a *Strategy for the Management and Disposal of Used Nuclear Fuel and High-Level Waste* in January 2013. The DOE strategy calls for a new nuclear waste management entity to develop consent-based storage and disposal sites, similar to recommendations by the BRC. No funding has been requested or appropriated since FY2011 to continue NRC licensing of the Yucca Mountain repository, although the House voted to provide DOE with $150.0 million and NRC $55.0 million for Yucca Mountain licensing activities in FY2015.

## MOST RECENT DEVELOPMENTS

Agreements for the first loan guarantees for nuclear power plants under the Energy Policy Act of 2005 were signed February 20, 2014, by Energy Secretary Ernest Moniz. The federal loan guarantees total $6.5 billion for two reactors being constructed at the Vogtle nuclear plant in Waynesboro, GA.[1] Southern Company, the plant's lead owner, estimated that its $3.46 billion share of the loan guarantee would reduce its present-value financing costs by up to $250 million. The guaranteed loans are to be issued by the U.S. Treasury's Federal Financing Bank.[2] The plant owners were not charged an upfront fee for the loan guarantees but are paying interest above the applicable Treasury rate.[3]

The Department of Energy's (DOE's) FY2015 funding request for nuclear energy research and development, submitted to Congress in early March 2014, totaled $863.4 million. Including advanced reactors, fuel cycle technology, infrastructure support, and safeguards and security, the total nuclear energy request is $25.0 million (2.9%) below the FY2014 funding level. The House approved its FY2015 Energy and Water appropriations bill (H.R. 4923) on July 10, 2014, with $826.7 million for nuclear energy. The Senate Subcommittee on Energy and Water Development Appropriations recommended $777.0 million on July 24, 2014.[4] Pending enactment of full-year appropriations, FY2015 funding for these programs is being provided by a continuing resolution (P.L. 113-164).

The first construction starts for new U.S. nuclear power reactors since the 1970s officially took place in March 2013 in South Carolina and Georgia. Pouring of the first "safety related" concrete, which marks the start of reactor construction, was completed on March 11, 2013, for V.C. Summer Unit 2 in Cayce, SC, and three days later for Vogtle Unit 3 in Georgia. The Nuclear Regulatory Commission (NRC) had issued combined construction permits and operating licenses (COLs) for two new reactors at the Vogtle site on February 9, 2012, and for two identical reactors at the Summer plant on March 30, 2012. Each of the new Westinghouse AP1000 reactors, scheduled for completion between 2017 and 2019, is expected to cost from $5 billion to $7 billion.

NRC approved a design certification September 16, 2014, for GE-Hitachi's Economic Simplified Boiling Water Reactor (ESBWR), following nine years of review. Like the Westinghouse AP1000, the ESBWR includes "passive" safety features that are designed to protect the reactor core from overheating after an accident. NRC is currently reviewing two COLs that would use the ESBWR design: Fermi 3 in Michigan and North Anna 3 in Virginia.[5]

Four U.S. reactors were permanently closed during 2013, and the shutdown of a fifth unit was announced for late 2014. Crystal River 3 in Florida was retired in February 2013 because of cracks in its concrete containment structure. The single-unit Kewaunee plant in Wisconsin closed in May 2013 because regional electricity prices had dropped below the reactor's generating costs. San Onofre 2 and 3 in California closed in June 2013 because of faulty steam generators (unit 1 had been shut previously). And the owner of the single-unit Vermont Yankee plant announced in August 2013 that the reactor would permanently close in the fourth quarter of 2014 for economic reasons. All of those units had substantial time remaining on their initial 40-year operating licenses or had received or applied for 20-year license extensions from NRC. The shutdowns prompted widespread discussion about the future of other aging U.S. reactors.

Continued operation of "at risk" nuclear generating capacity could be encouraged by regulations proposed by the Environmental Protection Agency (EPA) on June 2, 2014, to reduce carbon emissions at existing power plants. The EPA proposal would require states to reduce the "carbon intensity" (carbon emissions per megawatt-hour of electrical generation) of their existing fossil fuel-fired generating plants by 2030. The proposed standards assume that states will prevent the premature retirement of up to 6% of their existing nuclear capacity and also complete five reactors currently under construction. States could also respond to the standards by encouraging construction of additional reactors or the expansion of generating capacity at existing nuclear plants, among many other options.[6]

On March 12, 2012, NRC issued its first nuclear plant safety requirements based on lessons learned from the March 2011 Fukushima disaster in Japan. NRC ordered U.S. nuclear plant operators to begin implementing safety enhancements related to power blackouts, reactor containment venting, and monitoring the water levels of reactor spent fuel pools. The Fukushima nuclear plant was hit by an earthquake and tsunami that knocked out all electric power at the six-reactor plant, resulting in the overheating of the reactor cores in three of the units and a heightened overheating risk at several spent fuel storage pools at the site. The overheating of the reactor cores caused major hydrogen explosions and releases of radioactive material to the environment. NRC's response to the accident has been the subject of continuing congressional oversight.

After halting work on a long-planned nuclear waste repository at Yucca Mountain in Nevada, DOE issued a *Strategy for the Management and Disposal of Used Nuclear Fuel and High-Level Waste* in January 2013. The DOE strategy calls for a new nuclear waste management entity to develop consent-based storage and disposal sites. A pilot interim spent fuel storage facility would be opened by 2021 and a larger-scale storage facility, which could be an expansion of the pilot facility, would begin receiving waste by 2025. A geologic disposal facility would open by 2048— fifty years after the initial planned opening date for the Yucca Mountain repository.[7] Legislation to redirect the nuclear waste program along similar lines was introduced by Senator Wyden on June 27, 2013 (S. 1240). The U.S. Court of Appeals for the District of Columbia Circuit ruled on August 13, 2013, that NRC must continue work on the Yucca Mountain license application as long as funding is available. The Court determined that NRC has at least $11.1 million in previously appropriated funds for that purpose.[8] In its FY2015 Energy and Water Development appropriations bill, the House voted to provide DOE $150.0 million and NRC $55.0 million for Yucca Mountain licensing activities. No funding was included in the draft bill approved by the Senate subcommittee.

Pursuant to a court ruling, DOE stopped collecting nuclear waste fees from nuclear power generators on May 16, 2014, eliminating about $750 million in annual revenues for the waste disposal program.[9] The Nuclear Waste Policy Act requires the Secretary of Energy to adjust the fees as necessary to cover the waste program's anticipated costs, but the U.S. Court of Appeals for the District of Columbia Circuit ruled that DOE's current waste plans are too vague to allow a reasonable estimate to be calculated and that DOE must therefore stop collecting the fee.[10]

NRC approved a final rule August 26, 2014, on continued storage of spent nuclear fuel.[11] The rule responds to a federal circuit court ruling on June 8, 2012, that struck down NRC's Waste Confidence Decision, which contains the agency's formal findings that waste generated by nuclear power plants will be disposed of safely. The court ruled that the Waste Confidence Decision required an environmental review under the National Environmental Policy Act and that NRC needed to consider the possibility that a permanent waste repository would never be built and examine potential problems with waste storage pools. In approving the storage rule, NRC ended its suspension of final licensing decisions for new reactors, spent fuel storage facilities, and license renewals.[12]

## NUCLEAR POWER STATUS AND OUTLOOK

After nearly 30 years in which no new orders had been placed for nuclear power plants in the United States, a series of license applications that began in 2007 prompted widespread speculation about a U.S. "nuclear renaissance." The renewed interest in nuclear power largely resulted from the improved performance of existing reactors, federal incentives in the Energy Policy Act of 2005 (P.L. 109-58), the possibility of carbon dioxide controls that could increase costs at fossil fuel plants, and volatile prices for natural gas—the favored fuel for new power plants for the past two decades.

Four of the proposed new U.S. reactors received licenses from the Nuclear Regulatory Commission (NRC) in early 2012. NRC approved combined construction permit and operating licenses (COLs) for Southern Company to build and operate two new Westinghouse AP1000 reactors at the Vogtle nuclear power plant in Georgia on February 9, 2012. On March 30, 2012, NRC approved COLs for two additional AP1000 reactors at the existing V.C. Summer nuclear plant in South Carolina. Pouring of the first "safety related" concrete, which marks the start of reactor construction, was completed on March 11, 2013, for Summer Unit 2 and three days later for Vogtle Unit 3 (after several years of site preparation, component fabrication, and other preliminary work).

However, the future of all other proposed new U.S. reactors is uncertain. High construction cost estimates—a major reason for earlier reactor cancellations—continue to undermine nuclear power economics. A more recent obstacle to nuclear power growth has been the development of vast reserves of domestic natural gas from previously uneconomic shale formations, which has held gas prices low and reduced concern about future price spikes. Moreover, it is unclear at this point how much incentive might be provided to nuclear power by potential U.S. carbon controls.

Four U.S. reactors were permanently closed during 2013, and the shutdown of a fifth unit was announced for late 2014. Crystal River 3 in Florida was retired in February 2013 because of cracks in its concrete containment structure. The single-unit Kewaunee plant in Wisconsin closed in May 2013 because regional electricity prices had dropped below the reactor's generating costs. San Onofre 2 and 3 closed in June 2013 because of faulty steam generators (unit 1 had been shut previously). And the owner of the single-unit Vermont Yankee plant announced in August 2013 that the reactor would permanently close in the fourth quarter of 2014 for economic reasons. All of those units had substantial time remaining on their initial 40-year operating licenses or had received or applied for 20-year license extensions from NRC. The shutdowns prompted widespread discussion about the future of other aging U.S. reactors.

The March 11, 2011, earthquake and tsunami that severely damaged Japan's Fukushima Daiichi nuclear power plant could also affect plans for new U.S. reactors, although U.S. nuclear power growth was already expected to be modest in the near term. Following the Fukushima accident, preconstruction work was suspended on two planned reactors at the South Texas Project.

Tokyo Electric Power Company (TEPCO), which owns the Fukushima plant, had planned to invest in the South Texas Project expansion, but TEPCO's financial condition plunged after the accident. New U.S. safety requirements resulting from the Fukushima disaster could raise investor concerns about higher costs. On the other hand, after the accident the Obama Administration reiterated its support for nuclear power expansion as part of its clean energy policy.[13]

The recent applications for new power reactors in the United States followed a long period of declining nuclear generation growth rates. Until the COLs were issued for the Vogtle and Summer projects, no nuclear power plants had been ordered in the United States since 1978, and more than 100 reactors had been canceled, including all ordered after 1973. The most recent U.S. nuclear unit to be completed was the Tennessee Valley Authority's (TVA's) Watts Bar 1 reactor, ordered in 1970 and licensed to operate in 1996. But largely because of better operation and capacity expansion at existing reactors, annual U.S. nuclear generation has risen by about 20% since the startup of Watts Bar 1.[14] The U.S. nuclear power industry currently comprises 100 licensed reactors (excluding the four permanently closed in 2013) at 62 plant sites in 31 states and generates about 19% of the nation's electricity.[15] TVA's board of directors voted August 1, 2007, to resume construction on Watts Bar 2, which had been suspended in 1985; the renewed construction project was to cost about $2.5 billion and be completed in 2013. However, TVA announced on April 5, 2012, that completing Watts Bar 2 would cost up to $2 billion more than expected and take until 2015.[16] At TVA's request, NRC in March 2009 reinstated the construction authorization for the two-unit Bellefonte (AL) nuclear plant, which had been deferred in 1988 and canceled in 2006.[17] The TVA board voted on August 18, 2011, to complete construction of Bellefonte 1 after the Watts Bar 2 project is finished. Completing Bellefonte 1 was projected at that time to cost $4.9 billion, with operation to begin by 2020.[18] Citing lower electricity sales, TVA on June 12, 2013, announced sharp cutbacks at the Bellefonte site.[19]

Annual electricity production from U.S. nuclear power plants is much greater than that from oil and hydropower and other renewable energy sources. Nuclear generation has been overtaken by natural gas in recent years, and it remains well behind coal, which accounted for about 39% of U.S. electricity generation in 2013.[20] Nuclear plants generated more than half the electricity in four states in 2012—New Hampshire, New Jersey, South Carolina, and Vermont—and 12 states generated 25%-50% of their electricity from nuclear power.[21] The 789 billion net kilowatt-hours of nuclear electricity generated in the United States during 2013[22] was about the same as the nation's entire electrical output in the early 1960s, when the oldest of today's operating U.S. commercial reactors were ordered.[23]

Reasons for the 30-year halt in U.S. nuclear plant orders included high capital costs, public concern about nuclear safety and waste disposal, and regulatory compliance issues.

High construction costs may pose the most serious obstacle to nuclear power expansion. Construction costs for reactors completed since the mid-1980s ranged from $2 to $6 billion, averaging more than $3,900 per kilowatt of electric generating capacity (in 2011 dollars), far higher than commercial fossil fuel technologies. The nuclear industry predicts that new plant designs could be built for less than that if many identical plants were built in a series, but current estimates for new reactors show little if any reduction in cost.[24]

In contrast, average U.S. nuclear plant operating costs per kilowatt-hour dropped substantially since 1990, and expensive downtime has been steadily reduced. Licensed U.S. commercial reactors generated electricity at an average of 90% of their total capacity in 2013, according to the Energy Information Administration (EIA).[25]

Seventy-three commercial reactors have received 20-year license renewals from the Nuclear Regulatory Commission (NRC), giving them up to a total of 60 years of operation. License renewals for 19 additional reactors are currently under review, and more are anticipated, according to NRC.[26] However, as noted above, three reactors that have received license renewals, Vermont Yankee, Crystal River, and Kewaunee, are being permanently closed for economic reasons.

## Possible New Reactors

Electric utilities and other firms have announced plans to apply for COLs for more than 30 reactors (see **Table 1**).[27] (For a discussion of COLs, see the "Licensing and Regulation" section below.)

As noted above, construction is currently underway on four of the proposed new reactors, at the Vogtle and Summer sites. COLs are being actively pursued for 14 additional reactors (shown in **Table 1**), whose owners have not committed to actual construction but are keeping the option available if conditions are more favorable in the future. The experience of the first few reactors to be constructed is likely to be crucial in determining whether a wave of subsequent units will move forward as the nuclear industry envisions.

The two new Vogtle reactors are scheduled to go on line in 2017 and 2018,[28] the same years now planned for startup of the new Summer units,[29] although possible delays of more than a year have been reported.[30] EIA estimates that construction costs of new nuclear power plants will average $5,335 per kilowatt of capacity, or about $6.1 billion for an AP1000 unit, not including interest costs.[31] The two Summer units are expected to cost about $11.6 billion in 2012 dollars, according to regulatory filings,[32] while the two Vogtle units are projected by their primary owner to cost a total of $13.35 billion.[33]

Duke Energy's Levy County project, with two AP1000 units, is scheduled by NRC to receive a final decision on its COL in mid-2015, although Duke has terminated its engineering, procurement, and construction (EPC) contract for the project. Duke said it did not foresee a need for the plant as soon as previously planned, but "continues to regard the Levy site as a viable option for future nuclear generation."[34] COLs for six reactors at four other sites—Fermi (MI), South Texas Project, William States Lee (SC), and North Anna (VA)—are scheduled to be issued by 2016. The joint venture developing the proposed South Texas Plant reactors, Nuclear Innovation North America (NINA), is focusing solely on the COL and a DOE loan guarantee.[35]

The proposed new reactors at the Fermi and North Anna sites reached a milestone on September 16, 2014, when NRC approved the design certification for GE-Hitachi's Economic Simplified Boiling Water Reactor (ESBWR), following nine years of review. The COLs that are currently under review for both those units would use the ESBWR design. Like the Westinghouse AP1000, the ESBWR includes "passive" safety features that are designed to protect the reactor core from overheating after an accident.[36]

As shown in **Table 1**, the remaining three projects that have not suspended their COLs, with a total of four proposed reactors, do not have firm licensing schedules from NRC. As a result, these reactors appear unlikely to be completed before the early 2020s. Some of these proposed nuclear projects may require additional partners in order to proceed to construction, according to company announcements.[37]

Several other COL applications have been suspended, withdrawn, or shifted to early site permits (ESPs) only. Entergy suspended further license review of its planned GE ESBWR reactors at River Bend, LA, and Grand Gulf, MS, although it still has a previously issued ESP for Grand Gulf. AmerenUE suspended review of a COL for its proposed new Callaway unit in Missouri, and Exelon withdrew its COL application for a proposed two-unit plant in Victoria County, TX. Most recently, Duke Energy suspended its application for two new AP1000s at its Shearon Harris plant in North Carolina, and Luminant suspended its COL for two Mitsubishi APWRs at Comanche Peak in Texas.

TVA decided to defer consideration of its COL application for two new Westinghouse AP1000 reactors at its Bellefonte plant in Alabama in favor of completing the first of two unfinished Babcock & Wilcox reactors at the site. TVA had submitted a COL application for the Bellefonte AP1000s in October 2007 as part of the NuStart consortium.[38]

Constellation Energy announced October 9, 2010, that it was abandoning negotiations with DOE for a loan guarantee for the planned Calvert Cliffs 3 reactor, which Constellation had been developing as part of its UniStar joint venture with the French national utility EDF.[39] Constellation sold its share of UniStar to EDF so that EDF could seek another U.S. partner to continue the Calvert Cliffs project.[40] (For more discussion of Constellation's decision, see the "Loan Guarantees" section below.)

NRC anticipates that several more COL and other license applications will be submitted in the future. This includes a TVA plan to submit construction permit applications for small modular reactors (SMRs) of about 160 megawatts each at its Clinch River, TN, site.[41]

**Table 1. Announced Nuclear Plant License Applications**

| **Announced Applicant** | **Site** | **Reactor Type** | **Units** | **Status** |
|---|---|---|---|---|
| COL issued | | | | |
| Southern | Vogtle (GA) | Westinghouse AP1000 | 2 | COL application submitted 3/13/08; engineering, procurement, and construction (EPC) contract signed 4/8/08; ESP and limited construction approved 8/26/09; conditional DOE loan guarantee announced 2/16/10; NRC hearing held 9/27-28/11; COL approved 2/9/12; first "safety-related concrete" poured 3/14/13 |

| **Announced Applicant** | **Site** | **Reactor Type** | **Units** | **Status** |
|---|---|---|---|---|
| SCE&G | Summer (SC) | Westinghouse AP1000 | 2 | COL submitted 3/31/08; EPC contract signed 5/27/08; COL approved 3/30/12; first "safety-related concrete" poured 3/11/13 |
| COL scheduled for completion | | | | |
| Duke Energy | Levy County (FL) | Westinghouse AP1000 | 2 | COL submitted 7/30/08; application scheduled for completion in 2014; termination of EPC contract announced 8/1/13 |
| DTE Energy | Fermi (MI) | GE ESBWR | 1 | COL submitted 9/18/08; application scheduled for completion in 2015 |
| Nuclear Innovation North America | South Texas Project | Toshiba ABWR | 2 | COL submitted 9/20/07; EPC contract signed with Toshiba 2/12/09; NRG Energy halted further investment 4/19/11; application scheduled for completion in 2016 |
| Duke Energy | William States Lee (SC) | Westinghouse AP1000 | 2 | COL submitted 12/13/07; application scheduled for completion in 2016 |
| Dominion | North Anna | GE ESBWR | 1 | COL submitted 11/27/07; ESP approved 11/20/07; application scheduled for completion in 2016 |
| COL schedule under revision | | | | |
| FPL | Turkey Point (FL) | Westinghouse AP1000 | 2 | COL submitted 6/30/09; preconstruction work being conducted |
| PPL | Bell Bend (PA) | Areva EPR | 1 | COL submitted 10/10/08, not scheduled for review |
| UniStar | Calvert Cliffs (MD) | Areva EPR | 1 | COL submitted 7/13/07 (Part 1), 3/13/08 (Part 2); Constellation withdrew from project 10/8/10; application not scheduled for review |

**Table 1. (Continued)**

| Announced Applicant | Site | Reactor Type | Units | Status |
|---|---|---|---|---|
| Licensing suspended | | | | |
| Entergy | Grand Gulf (MS) | Not specified | 1 | COL submitted 2/27/08; licensing suspended 1/9/09; ESP approved 3/27/07 |
| Exelon | Victoria County (TX) | Not specified | 2 | COL application withdrawn and ESP application submitted 3/25/10; ESP application withdrawn 8/28/12 |
| AmerenUE | Calloway (MO) | Areva EPR | 1 | COL submitted 7/24/08; license review suspended 6/23/09 |
| Entergy | River Bend (LA) | Not specified | 1 | COL submitted 9/25/08; licensing suspended 1/9/09 |
| TVA | Bellefonte | Westinghouse AP1000 | 2 | COL submitted 10/30/07; licensing deferred 9/29/10 |
| Unistar | Nine Mile Point (NY) | Areva EPR | 1 | COL submitted 9/30/08; licensing suspended 12/1/09 |
| Duke Energy | Harris (NC) | Westinghouse AP1000 | 2 | COL submitted 2/19/08; EPC contract signed 1/5/09; licensing suspended 5/2/13 |
| Luminant Power | Comanche Peak (TX) | Mitsubishi US-APWR | 2 | COL submitted 9/19/08; review suspended 3/31/14 |
| Total COL applications | | 28 | | |
| Total currently active COLs | | 16 | | |

Sources: NRC, *Nucleonics Week*, *Nuclear News*, Nuclear Energy Institute, company news releases.

Note: Applications are for COLs unless otherwise specified.

# Nuclear Power Plant Safety and Regulation

## Safety

Worldwide concern about nuclear power plant safety rose sharply after the Fukushima accident, which is generally considered to be much worse than the

March 1979 Three Mile Island accident in Pennsylvania but not as severe as the April 1986 Chernobyl disaster in the former Soviet Union. Total radioactive releases from the Fukushima accident have been estimated at 25 million curies,[42] compared with 140 million curies from Chernobyl[43] and 43,000 curies from Three Mile Island.[44]

The Fukushima disaster resulted in similar levels of radioactive contamination per square meter to that of Chernobyl, but the Fukushima contamination was much less widespread and affected a smaller number of people.[45] Timely evacuation of areas up to 30 kilometers (km, 18.6 miles) from the Fukushima Daiichi plant reduced radiation exposure from the accident. A health survey conducted by the Fukushima prefecture estimated the total dose to the prefecture's 435,788 people at less than 617 sieverts (61,700 rem), averaging no more than about 1.4 millisieverts (140 millirem) per person for the first 28 months after the accident.[46] For comparison, NRC limits the dose resulting from the use of radioactive materials to 100 millirem per year (1 millisievert) to any individual member of the public.[47] (For more background on the Fukushima accident, see CRS Report R41694, *Fukushima Nuclear Disaster*, by Mark Holt, Richard J. Campbell, and Mary Beth D. Nikitin)[48]

The Fukushima disaster has raised particular policy questions for the United States because, unlike Chernobyl, the Fukushima reactors are similar to common U.S. designs. Although the Fukushima accident resulted from a huge tsunami that incapacitated the power plant's emergency diesel generators, the accident dramatically illustrated the potential consequences of any natural catastrophe or other situation that could cause an extended "station blackout" – the loss of alternating current (AC) power. Safety issues related to station blackout include standards for backup batteries, which had been required to provide power for 4-8 hours, and additional measures that may be required to assure backup power.

Safety concerns at U.S. reactors were also raised by hydrogen explosions at three of the Fukushima reactors—resulting from a high-temperature reaction between steam and nuclear fuel cladding—and the loss of cooling at the Japanese plant's spent fuel storage pools. Other safety issues that have been raised in the wake of Fukushima include the vulnerability of U.S. nuclear plants to earthquakes, floods, and other natural disasters, the availability of iodine pills to prevent absorption of radioactive iodine released during nuclear accidents, and the adequacy of nuclear accident emergency planning.

In response to such concerns, NRC on March 23, 2011, established a task force "made up of current senior managers and former NRC experts" to "conduct both short- and long-term analysis of the lessons that can be learned from the situation in Japan."[49] The Near-Term Task Force issued its report July 12, 2011, making recommendations ranging from specific safety improvements to broad changes in NRC's overall regulatory approach.[50] NRC staff subsequently identified several of those actions that "can and should be initiated without delay."[51] The NRC Commissioners largely agreed with the recommendations on October 18, 2011, and instructed the agency's staff to "strive to complete and implement the lessons learned from the Fukushima accident within five years—by 2016."[52] Tier 1 regulatory actions, which are now being implemented, include[53]

- *Seismic and flood hazard reevaluations and walkdowns.* Nuclear plant operators must evaluate the implications of updated seismic and flooding models, including all potential flooding sources. Plant operators must identify and verify the adequacy of flood and seismic protection features at their sites.
- *Station blackout regulatory actions.* NRC issued an order on March 12, 2012, that required U.S. reactors to implement mitigation strategies "that will allow them to cope without their permanent electrical power sources for an indefinite amount of time." Under the order, installed equipment at each plant must be sufficient to maintain or restore cooling until portable on-site equipment and supplies could take over. The portable on-site equipment would have to provide sufficient cooling until "sufficient offsite resources" could be brought in to maintain cooling indefinitely. Enough equipment and personnel would be required to protect all affected reactors at a multi-unit plant. NRC is currently preparing permanent regulations based on the mitigation strategies order.
- *Reliable hardened vents for Mark I containments.* NRC on March 12, 2012, ordered nuclear plants to install "reliable, hardened" vents for the containments in Mark I reactors (the type at Fukushima). The vents would be designed to reduce containment pressure before damage occurred to the reactor core. NRC modified the order in June 2013 to require that the vents continue to function after core damage occurred, which could prevent hydrogen generated by overheated fuel cladding from leaking into the reactor building, as occurred at Fukushima. Because venting after core damage has occurred could release radioactive core material into the environment, NRC is also

considering a requirement that vents include filters or that other strategies be implemented to reduce such emissions.

- *Spent fuel pool instrumentation.* NRC ordered nuclear plants on March 12, 2012, to install safety instrumentation to monitor spent fuel pool conditions, such as water level, temperature, and radiation levels, from the plant control room.
- *Strengthening and integrating accident procedures and capabilities.* NRC issued an Advanced Notice of Proposed Rulemaking on April 18, 2012, to require integrated emergency procedures, including clear command-and-control strategies and training qualifications for emergency decisionmakers.
- *Emergency preparedness regulatory actions.* NRC has asked nuclear plants how many workers would be needed to respond to large accidents affecting multiple reactors at the same site. In addition, plants were asked to assess and ensure the operability of emergency communications systems during such accidents.

The NRC staff slightly modified its proposals for top priority actions and divided the remaining Task Force proposals into two lower tiers, which were determined to require further assessment and potentially long-term study. Included in the lower-tier actions were requirements for emergency water supply systems for spent fuel pools, secure power for emergency communications and data systems, confirmation of seismic and flooding hazards, and modifications to NRC's regulatory process.[54]

### *Emergency Planning*

Following the Three Mile Island accident, which revealed severe weaknesses in preparations for nuclear plant emergencies, Congress mandated that emergency plans be prepared for all licensed power reactors (P.L. 96-295, Sec. 109). NRC was required to develop standards for emergency plans and review the adequacy of each plant-specific plan in consultation with the Federal Emergency Management Agency (FEMA).

NRC's emergency planning requirements focus on a "plume exposure pathway emergency planning zone (EPZ)," encompassing an area within about 10 miles (16.1 km) of each nuclear plant. Within the 10-mile EPZ, a range of responses must be developed to protect the public from radioactive releases, including evacuation, sheltering, and the distribution of non-radioactive iodine (as discussed above). The regulations also require a 50-mile "ingestion pathway EPZ," in which actions are developed to protect food supplies.[55]

Nuclear plants are required to conduct emergency preparedness exercises every two years. The exercises, which are evaluated by FEMA and NRC, may include local, state, and federal responders and may involve both the plume and ingestion EPZs.[56]

The size of the plume exposure EPZ has long been a subject of controversy, particularly after the 9/11 terrorist attacks on the United States, in which nuclear plants were believed to have been a potential target. Attention to the issue was renewed by the Fukushima accident, in which some of the highest radiation dose rates were measured beyond 10 miles from the plant.[57] Controversy over the issue intensified after NRC recommended the evacuation of U.S. citizens within 50 miles of the Fukushima plant on March 16, 2011. The NRC recommendation was based on computer models that, using meteorological data and estimates of plant conditions, found that potential radiation doses 50 miles from the plant could exceed U.S. protective action guidelines.[58]

In response to the 9/11 terrorist attacks, NRC modified its nuclear plant emergency planning requirements and began a comprehensive review of emergency planning regulations and guidance. An NRC final rule based on that review took effect December 23, 2011.[59] Among the changes included in the rule are new requirements for periodic updates of EPZ evacuation time estimates, mandatory backups for public alert systems, and protection of emergency responders during terrorist attacks. The new emergency planning regulations were prepared before the Fukushima accident, but the NRC staff recommended approval of the changes without waiting for further changes that might result from the lessons of the Japanese accident. Emergency planning changes resulting from Fukushima should be implemented later, the staff recommended.[60]

### *Domestic Reactor Safety Experience*

Nuclear power safety has been a longstanding issue in the United States. Safety-related shortcomings have been identified in the construction quality of some plants, plant operation and maintenance, equipment reliability, emergency planning, and other areas. In one serious case, it was discovered in March 2002 that leaking boric acid had eaten a large cavity in the top of the reactor vessel in Ohio's Davis-Besse nuclear plant. The corrosion left only the vessel's quarterinch-thick stainless steel inner liner to prevent a potentially catastrophic loss of reactor cooling water. Davis-Besse remained closed for repairs and other safety improvements until NRC allowed the reactor to restart in March 2004.

NRC's oversight of the nuclear industry is a subject of contention as well; nuclear utilities often complain that they are subject to overly rigorous and inflexible regulation, but nuclear critics charge that NRC frequently relaxes safety standards when compliance may prove difficult or costly to the industry.

In terms of public health consequences, the safety record of the U.S. nuclear power industry in comparison with other major commercial energy technologies has been excellent. During more than 3,500 reactor-years of operation in the United States,[61] the only incident at a commercial nuclear power plant that might lead to any deaths or injuries to the public has been the Three Mile Island accident, in which more than half the reactor core melted.[62] A study of 32,000 people living within five miles of the reactor when the accident occurred found no significant increase in cancer rates through 1998, although the authors noted that some potential health effects "cannot be definitively excluded."[63]

The relatively small amounts of radioactivity released by nuclear plants during normal operation are not generally believed to pose significant hazards, although some groups contend that routine emissions are unacceptably risky. There is substantial scientific uncertainty about the level of risk posed by low levels of radiation exposure; as with many carcinogens and other hazardous substances, health effects can be clearly measured only at relatively high exposure levels. In the case of radiation, the assumed risk of low-level exposure has been extrapolated mostly from health effects documented among persons exposed to high levels of radiation, particularly Japanese survivors of nuclear bombing in World War II, medical patients, and nuclear industry workers.[64]

NRC announced April 7, 2010, that it had asked the National Academy of Sciences (NAS) to "perform a state-of-the-art study on cancer risk for populations surrounding nuclear power facilities." Unlike in previous studies, NAS is to examine cancer diagnosis rates, rather than cancer deaths, potentially increasing the amount of data. The new study would also use geographic units smaller than counties to determine how far members of the study group are located from reactors, to more clearly determine whether there is a correlation between cancer cases and distance from reactors. After issuing a report on methodologies for the study in 2012, NAS began planning a pilot study at seven nuclear facilities in late 2013.[65]

NRC's 1986 Safety Goal Policy Statement declared that nuclear power plants should not increase the risk of accidental or cancer deaths among the nearby population by more than 0.1%.[66] Later NRC guidance established a "subsidiary benchmark" for the probability of accidental core damage (fuel melting): Core damage frequency should average no more than one in 10,000 per reactor per year.[67] In addition, NRC set a benchmark that reactor containments should be successful at least 90% of the time in preventing major radioactive releases during a core-damage accident. Therefore, the benchmark probability of a major release from containment failure during a core melt accident would average less than one in 100,000 per reactor per year.[68] (For the current U.S. fleet of about 100 reactors, that rate would yield an average of one core-damage accident every 100 years and a major release every 1,000 years.) On the other hand, some groups challenge the complex calculations that go into predicting such accident frequencies, contending that accidents with serious public health consequences may be more frequent.[69]

### *Reactor Safety in the Former Soviet Bloc*

The Chernobyl accident was by far the worst nuclear power plant accident to have occurred anywhere in the world. At least 31 persons died quickly from acute radiation exposure or other injuries, and thousands of additional cancer deaths among the tens of millions of people exposed to radiation from the accident may occur during the next several decades.

According to a 2006 report by the Chernobyl Forum organized by the International Atomic Energy Agency, the primary observable health consequence of the accident was a dramatic increase in childhood thyroid cancer. The Chernobyl Forum estimated that about 4,000 cases of thyroid cancer have occurred in children who after the accident drank milk contaminated with high levels of radioactive iodine, which concentrates in the thyroid. Although the Chernobyl Forum found only 15 deaths from those thyroid cancers, it estimated that about 4,000 other cancer deaths may have occurred among the 600,000 people with the highest radiation exposures, plus an estimated 1% increase in cancer deaths among persons with less exposure. The report estimated that about 77,000 square miles were significantly contaminated by radioactive cesium.[70] Greenpeace issued a report in 2006 estimating that 200,000 deaths in Belarus, Russia, and Ukraine resulted from the Chernobyl accident between 1990 and 2004.[71]

## Licensing and Regulation

For many years, a top priority of the U.S. nuclear industry was to modify the process for licensing new nuclear plants. No electric utility would consider ordering a nuclear power plant, according to the industry, unless licensing became quicker and more predictable, and designs were less subject to mid-construction safety-related changes required by NRC. The Energy Policy Act of 1992 (P.L. 102-486) largely implemented the industry's licensing goals.

Nuclear plant licensing under the Atomic Energy Act of 1954 (P.L. 83-703; U.S.C. 2011-2282) had historically been a two-stage process. NRC first issued a construction permit to build a plant and then, after construction was finished, an operating license to run it. Each stage of the licensing process involved adjudicatory proceedings. Environmental impact statements also are required under the National Environmental Policy Act.

Over the vehement objections of nuclear opponents, the Energy Policy Act of 1992 provided a clear statutory basis for one-step nuclear licenses. Under the new process, NRC can issue combined construction permits and operating licenses (COLs) and allow completed plants to operate without delay if they meet all construction requirements—called "inspections, tests, analyses, and acceptance criteria," or ITAAC. NRC would hold preoperational hearings on the adequacy of plant construction only in specified circumstances.

DOE's Nuclear Power 2010 program had paid up to half the cost of several COLs and early site permits to test the revised licensing procedures. However, the COL process cannot be fully tested until construction of new reactors is completed. At that point, it could be seen whether completed plants would be able to operate without delays or whether adjudicable disputes over construction adequacy may arise. Section 638 of the Energy Policy Act of 2005 (EPACT05, P.L. 109-58) authorizes federal payments to the owner of a completed reactor whose operation is held up by regulatory delays. The nuclear industry has asked Congress to require NRC to use informal procedures in determining whether ITAAC have been met, eliminate mandatory hearings on uncontested issues before granting a COL, and make other changes in the licensing process.[72]

A fundamental concern in the nuclear regulatory debate is the performance of NRC in issuing and enforcing nuclear safety regulations. The nuclear industry and its supporters have regularly complained that unnecessarily stringent and inflexibly enforced nuclear safety regulations have burdened nuclear utilities and their customers with excessive costs. But many environmentalists, nuclear opponents, and other groups charge NRC with being too close to the nuclear industry, a situation that they say has resulted in lax oversight of nuclear power plants and routine exemptions from safety requirements.

Primary responsibility for nuclear safety compliance lies with nuclear plant owners, who are required to find any problems with their plants and report them to NRC. Compliance is also monitored directly by NRC, which maintains at least two resident inspectors at each nuclear power plant. The resident inspectors routinely examine plant systems, observe the performance of reactor personnel, and prepare regular inspection reports. For serious safety violations, NRC often dispatches special inspection teams to plant sites.

NRC's reactor safety program is based on "risk-informed regulation," in which safety enforcement is guided by the relative risks identified by detailed individual plant studies. NRC's risk-informed reactor oversight system, inaugurated April 2, 2000, relies on a series of performance indicators to determine the level of scrutiny that each reactor should receive.[73]

## Reactor Security

Nuclear power plants have long been recognized as potential targets of terrorist attacks, and critics have long questioned the adequacy of requirements for nuclear plant operators to defend against such attacks. All commercial nuclear power plants licensed by NRC have a series of physical barriers against access to vital reactor areas and are required to maintain a trained security force to protect them.

A key element in protecting nuclear plants is the requirement that simulated terrorist attacks, monitored by NRC, be carried out to test the ability of the plant operator to defend against them. The severity of attacks that plant security must prepare for is specified in the "design basis threat" (DBT).

EPACT05 required NRC to revise the DBT based on an assessment of terrorist threats, the potential for multiple coordinated attacks, possible suicide attacks, and other criteria. NRC approved the DBT revision based on those requirements on January 29, 2007. The revised DBT does not require nuclear power plants to defend against deliberate aircraft attacks. NRC contended that nuclear facilities were already required to mitigate the effects of large fires and explosions, no matter what the cause, and that active protection against airborne threats was being addressed by U.S. military and other agencies.[74] After much consideration, NRC voted February 17, 2009, to require all new nuclear power plants to incorporate design features that would ensure that, in the event of a crash by a large commercial aircraft, the reactor core would remain cooled or the reactor containment would remain intact, and radioactive releases would not occur from spent fuel storage pools.[75] The rule change was published in the Federal Register June 12, 2009.[76]

NRC rejected proposals that existing reactors also be required to protect against aircraft crashes, such as by adding large external steel barriers. However, NRC did impose some additional requirements related to aircraft crashes on all reactors, both new and existing, after the 9/11 terrorist attacks of 2001. In 2002, as noted above, NRC ordered all nuclear power plants to develop strategies to mitigate the effects of large fires and explosions that could result from aircraft crashes or other causes. An NRC regulation on fire mitigation strategies, along with requirements that reactors establish procedures for responding to specific aircraft threats, was approved December 17, 2008.[77] The fire mitigation rules were published in the Federal Register March 27, 2009.[78]

Other ongoing nuclear plant security issues include the vulnerability of spent fuel pools, which hold highly radioactive nuclear fuel after its removal from the reactor, standards for nuclear plant security personnel, and nuclear plant emergency planning. NRC's March 2009 security regulations addressed some of those concerns and included a number of other security enhancements.

EPACT05 required NRC to conduct force-on-force security exercises at nuclear power plants every three years (which was NRC's previous policy), authorized firearms use by nuclear security personnel (preempting some state restrictions), established federal security coordinators, and required fingerprinting of nuclear facility workers.

## Decommissioning

When nuclear power plants reach the end of their useful lives, they must be safely removed from service, a process called *decommissioning*. NRC requires nuclear utilities to make regular contributions to dedicated funds to ensure that money is available to remove radioactive material and contamination from reactor sites after they are closed.

The first full-sized U.S. commercial reactors to be decommissioned were the Trojan plant in Oregon, whose decommissioning completion received NRC approval on May 23, 2005, and the Maine Yankee plant, for which NRC approved most of the site cleanup on October 3, 2005. The Trojan decommissioning cost $429 million, according to reactor owner Portland General Electric, and the Maine Yankee decommissioning cost about $500 million.[79] Decommissioning of the Connecticut Yankee plant cost $790 million and was approved by NRC on November 26, 2007.[80] NRC approved the cleanup of the decommissioned Rancho Seco reactor site in California on October 7, 2009.[81] The decommissioning of Rancho Seco was estimated to cost $500 million, excluding future demolition of the cooling towers and other remaining plant structures.[82] Total costs for decommissioning, spent fuel storage, and site restoration at the recently closed San Onofre units 2 and 3 are estimated at $4.1 billion.[83] When a reactor is permanently shut down, the owner (licensee) has 30 days to notify NRC. The licensee then certifies with NRC when spent fuel has been permanently removed from the reactor vessel. By two years after shutdown, the licensee must submit a Post Shutdown Decommissioning Activities Report (PSDAR). The PSDAR specifies which of the two primary decommissioning options will be pursued:

- DECON: Plant and equipment are dismantled and removed, or decontaminated to the level required for release from NRC licensing.
- SAFSTOR: The plant is placed in a safe, stable condition for future dismantlement and decontamination.

According to NRC, nine reactors are currently in SAFSTOR: Dresden 1 (IL), Indian Point 1 (NY), La Crosse (WI), Millstone 1 (CT), Peach Bottom 1 (PA), San Onofre 1 (CA), GE Vallecitios (CA), NS Savannah (MD), and Three Mile Island 2 (PA). Four units are in DECON: Fermi 1 (MI), Humboldt Bay (CA), and Zion 1 and 2 (IL).[84]

After nuclear reactors are decommissioned, the spent nuclear fuel (SNF) accumulated during their operating lives remains stored in pools or dry casks at the plant sites. About 5,900 metric tons of spent fuel is currently stored at 12 closed nuclear power plants and Vermont Yankee, which is to close in late 2014.[85] "Until this SNF is removed from these [at the time] nine sites, the sites cannot be fully decommissioned and made available for other purposes," DOE noted in a 2008 report.[86] President Obama's decision to terminate development of an underground spent fuel repository at Yucca Mountain, NV, has increased concerns about the ultimate disposition of spent fuel at decommissioned sites.

## Nuclear Accident Liability

Liability for damages to the general public from nuclear incidents is addressed by the Price-Anderson Act (primarily Section 170 of the Atomic

Energy Act of 1954, 42 U.S.C. 2210). EPACT05 extended the availability of Price-Anderson coverage for new reactors and new DOE nuclear contracts through the end of 2025. (Existing reactors and contracts were already covered.)

Under Price-Anderson, the owners of commercial reactors must assume all liability for nuclear damages awarded to the public by the court system, and they must waive most of their legal defenses following a severe radioactive release ("extraordinary nuclear occurrence"). To pay any such damages, each licensed reactor with at least 100 megawatts of electric generating capacity must carry the maximum liability insurance reasonably available, which was raised from $300 million to $375 million on January 1, 2010.[87] Any damages exceeding $375 million are to be assessed equally against all 100-megawatt-and-above power reactors, up to $121.3 million per reactor (increased for inflation from $111.9 million on September 10, 2013).[88] Those assessments—called "retrospective premiums"—would be paid at an annual rate of no more than $19.0 million per reactor (up from $17.5 million), to limit the potential financial burden on reactor owners following a major accident. According to NRC, 104 commercial reactors, including the four closed in 2013,[89] are currently covered by the Price-Anderson retrospective premium requirement.[90]

For each nuclear incident, the Price-Anderson liability system currently would provide up to $13.6 billion in public compensation. That total includes $121.3 million in retrospective premiums from each of the 104 currently covered reactors, totaling $12.6 billion, plus the $375 million in insurance coverage carried by the reactor that suffered the incident. On top of those payments, a 5% surcharge may also be imposed, raising the total per-reactor retrospective premium to $127.4 million and the total available compensation to about $13.6 billion. Under Price-Anderson, the nuclear industry's liability for an incident is capped at that amount, which varies over time depending on the number of covered reactors, the amount of available insurance, and the inflation adjustment. Payment of any damages above that liability limit would require congressional approval under special procedures in the act.

EPACT05 increased the limit on per-reactor annual payments to $15 million from the previous $10 million, and required the annual limit to be adjusted for inflation every five years. As under previous law, the total retrospective premium limit is adjusted every five years as well. For the purposes of those payment limits, a nuclear plant consisting of multiple small reactors (100-300 megawatts, up to a total of 1,300 megawatts) would be considered a single reactor. Therefore, in the event of a severe release a power plant with six 120-megawatt small modular reactors would be liable for retrospective premiums of up to $121.3 million, rather than $727.8 million (excluding the 5% surcharge).

The Price-Anderson Act also covers contractors who operate DOE nuclear facilities. EPACT05 set the liability limit on DOE contractors at $10 billion per accident, to be adjusted for inflation every five years. The first adjustment under EPACT, raising the liability limit to $11.961 billion, took effect October 14, 2009.[91] The liability limit for DOE contractors previously had been the same as for commercial reactors, excluding the 5% surcharge, except when the limit for commercial reactors dropped because of a decline in the number of covered reactors. Price-Anderson authorizes DOE to indemnify its contractors for the entire amount of their liability, so that damage payments for nuclear incidents at DOE facilities would ultimately come from the Treasury. However, the law also allows DOE to fine its contractors for safety violations, and contractor employees and directors can face criminal penalties for "knowingly and willfully" violating nuclear safety rules. EPACT05 limited the civil penalties against a nonprofit contractor to the amount of management fees paid under that contract.

The Price-Anderson Act's limits on liability were crucial in establishing the commercial nuclear power industry in the 1950s. Supporters of the Price-Anderson system contend that it has worked well since that time in ensuring that nuclear accident victims would have a secure source of compensation, at little cost to the taxpayer. Extension of the act was widely considered a prerequisite for new nuclear reactor construction in the United States. Opponents contend that Price-Anderson inappropriately subsidizes the nuclear power industry by reducing its insurance costs and protecting it from some of the financial consequences of the most severe conceivable accidents. Projections that damages to the public from the Fukushima accident will greatly exceed the Price-Anderson liability limits prompted renewed calls for reexamination of the law.[92]

The U.S. government is supporting the establishment of an international liability system that, among other purposes, would cover U.S. nuclear equipment suppliers conducting foreign business. The Convention on Supplementary Compensation for Nuclear Damage (CSC) will not enter into force until at least five countries with a specified level of installed nuclear capacity have enacted implementing legislation. Such implementing language was included in the Energy Independence and Security Act of 2007 (P.L. 110-140, section 934), signed by President Bush December 19, 2007. Supporters of the Convention hope that more countries will join now that the United States has acted. Aside from the United States, four countries have submitted the necessary instruments of ratification, but they do not have the required nuclear capacity for it to take effect.[93] Canada signed the convention on December 3, 2013, with the government announcing that it would be submitted to Parliament for approval. Ratification by Canada, which has a large nuclear reactor fleet, would put the convention into force 90 days later.[94]

Under the U.S. implementing legislation, the CSC would not change the liability and payment levels already established by the Price-Anderson Act. Each party to the convention would be required to establish a nuclear damage compensation system within its borders analogous to Price-Anderson. For any damages not covered by those national compensation systems, the convention would establish a supplemental tier of damage compensation to be paid by all parties. P.L. 110-140 requires the U.S. contribution to the supplemental tier to be paid by suppliers of nuclear equipment and services, under a formula to be developed by DOE. Supporters of the convention contend that it will help U.S. exporters of nuclear technology by establishing a predictable international liability system. For example, U.S. nuclear equipment sales to the growing economies of China and India would be facilitated by those countries' participation in the CSC liability regime.

## FEDERAL INCENTIVES FOR NEW NUCLEAR PLANTS

The nuclear power industry contends that support from the federal government would be needed for "a major expansion of nuclear energy generation."[95] Significant incentives for building new nuclear power plants were included in the Energy Policy Act of 2005 (EPACT05, P.L. 109-58), signed by President Bush on August 8, 2005. These include production tax credits, loan guarantees, insurance against regulatory delays, and extension of the Price-Anderson Act nuclear liability system (discussed in the previous section on "Nuclear Accident Liability"). Relatively low prices for natural gas—nuclear power's chief competitor—and rising estimated nuclear plant construction costs have decreased the likelihood that new reactors would be built without federal support. Any regulatory delays and increased safety requirements resulting from the Fukushima accident could also pose an obstacle to nuclear construction plans.

As a result, numerous bills have been introduced in recent years to strengthen or add to the EPACT05 incentives (see "Legislation in the 113$^{th}$ Congress" at the end of this report). Nuclear power critics have denounced the federal support programs and proposals as a "bailout" of the nuclear industry, contending that federal efforts should focus instead on renewable energy and energy efficiency.[96]

## Nuclear Production Tax Credit

EPACT05 provides a 1.8-cents/kilowatt-hour tax credit for up to 6,000 megawatts of new nuclear capacity for the first eight years of operation, up to $125 million annually per 1,000 megawatts. The credit is not adjusted for inflation.

The Treasury Department published interim guidance for the nuclear production tax credit on May 1, 2006.[97] Under the guidance, the 6,000 megawatts of eligible capacity (enough for about four or five reactors) are to be allocated among reactors that filed license applications by the end of 2008. If more than 6,000 megawatts of nuclear capacity ultimately qualify for the production tax credit, then the credit is to be allocated proportionally among any of the qualifying reactors that begin operating before 2021.

By the end of 2008, license applications had been submitted to NRC for more than 34,000 megawatts of nuclear generating capacity,[98] so if all those reactors were built before 2021 they would receive less than 20% of the maximum tax credit. However, the reactor licensing status shown in **Table 1** indicates that only four new units, totaling about 4,600 megawatts of capacity, are currently licensed for construction and likely to be completed before 2021. Eight other units, totaling about 10,300 megawatts, are scheduled to receive their licenses by 2016 and could possibly go into service by 2021.

The Nuclear Energy Institute (NEI) has urged Congress to remove the 6,000 megawatt capacity limit for the production tax credit, index it for inflation, and extend the deadline for plants to begin operation to the start of 2025. NEI also proposed that a 30% investment tax credit be available for new nuclear construction as an alternative to the production credit.[99]

## Standby Support

Because the nuclear industry has often blamed licensing delays for past nuclear reactor construction cost overruns, EPACT05 authorizes the Secretary of Energy to provide "standby support," or regulatory risk insurance, to help pay the cost of regulatory delays at up to six new commercial nuclear reactors. For the first two reactors that begin construction, the DOE payments could cover all the eligible delay-related costs, such as additional interest, up to $500 million each. For the next four reactors, half of the eligible costs could be paid by DOE, with a payment cap of $250 million per reactor. Delays caused by the failure of a reactor owner to comply with laws or regulations would not be covered. Project sponsors will be required to pay the "subsidy cost" of the program, consisting of the estimated present value of likely future government payments. DOE published a final rule for the "standby support" program August 11, 2006.[100]

Under the program's regulations, a project sponsor may enter into a conditional agreement for standby support before NRC issues a combined operating license. The first six conditional agreements to meet all the program requirements, including the issuance of a COL and payment of the estimated subsidy costs, can be converted to standby support contracts. However, no applicant has pursued the incentive.[101]

## Loan Guarantees

Title XVII of EPACT05 authorizes federal loan guarantees for up to 80% of construction costs for advanced energy projects that reduce greenhouse gas emissions, including new nuclear power plants. Under such loan guarantee agreements, the federal government would repay all covered loans if the borrower defaulted. This would reduce the risk to lenders and allow them to provide financing at low interest rates. The Title XVII loan guarantees are widely considered crucial by the nuclear industry to obtain financing for new reactors. However, opponents contend that nuclear loan guarantees provide an unjustifiable subsidy to a mature industry and shift investment away from environmentally preferable energy technologies.[102] The authorized ceiling on nuclear power plant loan guarantees is currently $18.5 billion.[103]

Agreements for the first loan guarantees for nuclear power plants under EPACT05 were signed February 20, 2014, by Energy Secretary Ernest Moniz. The federal loan guarantees total $6.5 billion for the two reactors being constructed at the Vogtle nuclear plant in Georgia—$3.46 billion for Southern Company, the plant's lead owner, and $3.06 billion for Oglethorpe Power Corporation. A further $1.8 billion loan guarantee for the Vogtle project, for the Municipal Electric Authority of Georgia, has not been finalized. Southern Company estimated that its Vogtle loan guarantees would reduce its present-value financing costs by up to $250 million.[104]

***Subsidy Costs***

Title XVII requires the estimated future government costs resulting from defaults on guaranteed loans to be covered up-front by appropriations or by payments from project sponsors, such as the utility planning to build a plant. As specified by the Federal Credit Reform Act of 1990 (FCRA, contained in P.L. 101-508), these "subsidy costs" are calculated as the present value of the average possible future net costs to the government for each loan guarantee. If those calculations are accurate, the subsidy cost payments for all the guaranteed projects together should cover the future costs of the program, including default-related losses. However, the Congressional Budget Office has predicted that the up-front subsidy cost payments will prove too low by at least 1% and is scoring bills accordingly.[105] For example, appropriations bills that provide loan guarantee authorizations include an adjustment equal to 1% of the loan guarantee ceiling.

The guaranteed loans for the new Vogtle reactors are to be issued by the U.S. Treasury's Federal Financing Bank. No subsidy cost was charged for the Vogtle loan guarantees, which raised considerable comment, as the risk of the project is likely to be greater than zero. The methodology that may have led to such a result is discussed below.

In order to estimate credit subsidy costs for a loan transaction, certain information is required: (1) loan value, (2) loan term (years), (3) credit risk, including default rates and recovery rates, (4) loan interest rate, and (5) Treasury interest rate(s). Details of the credit subsidy cost calculations for each loan (i.e., default rates, recovery rates, and project credit rating) have not been made public. However, some aspects and considerations of the Vogtle project loans may provide some perspective regarding the zero credit subsidy fee assessment.

Defaults—net of recoveries—and interest are the two primary subsidy cost components that likely resulted in the zero subsidy fee calculation. The default subsidy represents the present value of expected government losses due to defaults less recoveries. Default rates and recovery rates—based on loan credit risk and other parameters—are applied to the outstanding principal balance for each payment period. Two aspects of the Vogtle loans may have contributed to either low default rates and/or high recovery rates: (1) strong financial condition and obligations of the borrowers, and (2) the possibility of costs being passed through to Georgia Power and Oglethorpe electricity consumers, thereby resulting in low repayment risk, and therefore a low default subsidy estimate. These characteristics raise questions about the need for federal credit support for this project. Nevertheless, it is likely that default-related losses to the government exist.

Interest rates for the Vogtle project loans range from 3.5% to 3.9%—higher than Treasury rates— for disbursements made in February 2014. It is possible that gains to the government from higher interest rates—relative to Treasury securities—offset any estimated default-related losses to the government, while also taking into account other payments from the borrower or costs to the government. Should this be the case, the FCRA-prescribed calculation methodology could result in a zero, or even a negative, credit subsidy cost.

The nuclear industry contends that historical experience indicates defaults are likely to be minimal and that nuclear plant subsidy costs should therefore be low.[106] However, nuclear power critics contend that nuclear power plants are likely to experience delays and cost overruns that could lead to much larger losses under the loan guarantee program. The Center for American Progress concluded that nuclear subsidy costs "should be at least 10 percent and possibly much more."[107]

Constellation Energy informed DOE on October 8, 2010, that it was withdrawing from loan guarantee negotiations on Calvert Cliffs 3, blaming "the Office of Management and Budget's inability to address significant problems with its methodology for determining the project's credit subsidy cost." Constellation's letter to DOE said OMB's "shockingly high" estimate of the subsidy cost for Calvert Cliffs 3 was 11.6%, or about $880 million. "Such a sum would clearly destroy the project's economics (or the economics of any nuclear project for that matter), and was dramatically out of line with both our own and independent assessments of what the figure should reasonably be," the letter stated.[108] Although OMB has not released its subsidy cost methodology, it may consider the default risk for a "merchant plant" such as Calvert Cliffs to be significantly higher than that of a rate-regulated plant such as Vogtle. A plant under traditional rate regulation is allowed to pass all prudently incurred costs through to utility ratepayers, while a merchant plant charges market rates for its power. A merchant plant, therefore, could potentially earn higher profits than a rate-regulated plant, but it also runs the risk of being unable to cover its debt payments if market rates for wholesale electric power drop too low or if its costs are higher than anticipated.

***Nuclear Solicitations***

DOE issued a solicitation for up to $20.5 billion in nuclear power and uranium enrichment plant loan guarantees on June 30, 2008.[109] According to the nuclear industry, 10 nuclear power projects applied for $93.2 billion in loan guarantees, and two uranium enrichment projects asked for $4.8 billion in guarantees, several times the amount available.[110]

In the uranium enrichment solicitation, DOE in July 2009 informed USEC Inc., which proposes to build a new plant in Ohio, that its technology needed further testing before a loan guarantee could be issued.[111] DOE notified Congress in March 2010 that it would reprogram $2 billion of its unused FY2007 loan guarantee authority toward uranium enrichment, increasing the uranium enrichment total to $4 billion. The move would potentially allow guarantees to be provided to both USEC and the other applicant in the uranium enrichment solicitation, the French firm Areva, which has proposed a plant in Idaho.[112] DOE offered a $2 billion conditional loan guarantee to Areva on May 20, 2010.[113]

DOE informed USEC in October 2011 that the centrifuge technology for its proposed new enrichment plant still needed further testing and offered to provide up to $300 million to help build a demonstration "train" of 720 centrifuges.[114] The FY2013 Continuing Appropriations Resolution (P.L. 112-175) included $100 million for the USEC demonstration program.[115] The FY2014 Consolidated Appropriations Act (P.L. 113-76, Sec. 321) provided up to $56 million for the program through special reprogramming authority.

DOE has recently provided other assistance to USEC. DOE agreed on May 15, 2012, to provide depleted uranium stockpiles (material left over from the enrichment process) to Energy Northwest for reenrichment at USEC's plant in Paducah, KY, for use as reactor fuel.[116] DOE agreed on March 13, 2012, to acquire low-enriched uranium from USEC in exchange for taking responsibility for low-value depleted uranium tails that USEC would otherwise have to dispose of, freeing $44 million of USEC's funds for the centrifuge project.[117] DOE announced June 13, 2012, that it would provide $88 million for the centrifuge demonstration program by taking over responsibility for disposal of additional depleted uranium from USEC. In return, DOE will take ownership of the equipment and technology used in the demonstration and lease it to USEC.[118]

Despite the DOE assistance, USEC filed for bankruptcy reorganization March 5, 2014. The U.S. Bankruptcy Court for the District of Delaware approved USEC's reorganization plan September 5, 2014, in which the company will be renamed Centrus Energy Corporation.[119] DOE's Oak Ridge National Laboratory took over management of USEC's centrifuge demonstration project in April 2014, making USEC a subcontractor.[120]

DOE issued a draft solicitation for an additional $12.6 billion in nuclear loan guarantees on September 30, 2014, including new nuclear reactors, upgrades and uprates of existing nuclear reactors, and nuclear fuel facilities.[121]

## Global Climate Change

Global climate change that may be caused by carbon dioxide and other greenhouse gas emissions is cited by nuclear power supporters as an important reason to develop a new generation of reactors. Nuclear power plants emit relatively little carbon dioxide, mostly from nuclear fuel production and auxiliary plant equipment. This "green" nuclear power argument has received growing attention in think tanks and academia. As stated by the Massachusetts Institute of Technology in its major study *The Future of Nuclear Power*: "Our position is that the prospect of global climate change from greenhouse gas emissions and the adverse consequences that flow from these emissions is the principal justification for government support of the nuclear energy option."[122] The Obama Administration is including nuclear power as part of its clean energy strategy.

However, some environmental groups have contended that nuclear power's potential greenhouse gas benefits are modest and must be weighed against the technology's safety risks, its potential for nuclear weapons proliferation, and the hazards of radioactive waste.[123] They also contend that energy efficiency and renewable energy would be far more productive investments for reducing greenhouse gas emissions.[124]

Proposals to reduce carbon dioxide emissions – through taxation, a cap-and-trade system, or other regulatory controls – could significantly increase the cost of generating electricity with fossil fuels and improve the competitive position of nuclear power. A federal Clean Energy Standard that includes nuclear power, as proposed in President Obama's January 2011 State of the Union Address, could provide a similar boost to nuclear energy expansion. Utilities that have applied for nuclear power plant licenses have often cited the possibility of federal greenhouse gas controls or other mandates as one of the reasons for pursuing new reactors.

The Environmental Protection Agency (EPA) published a proposed rule on June 18, 2014, to address $CO_2$ emissions from existing power plants.[125] The potential effect of the proposed rule on nuclear power has become a significant policy question. The formula in the proposed rule for setting state $CO_2$ goals explicitly accounts for some existing nuclear capacity and reactors under construction, providing a potential incentive for states to try to keep those plants operating. However, EPA's proposed rule allows states to develop their own plans for meeting the $CO_2$ emission rate goals, making it difficult to predict how nuclear energy might ultimately fare.

The proposed EPA standards would set state-specific goals for the amount of $CO_2$ that could be emitted in 2030 for each megawatt-hour (MWh) of electricity generated. EPA projects that, under those proposed emissions rates, U.S. power plants would produce 30% less $CO_2$ by 2030 than they did in 2005 (the base year in President Obama's Climate Action Plan).[126]

The unadjusted baseline for the proposed standards was developed by dividing 2012 power plant $CO_2$ emissions in each state by the amount of electricity generated by fossil fuel-fired plants. In the case of Ohio, which EPA provided as an example, the unadjusted $CO_2$ emission rate for fossil plants in 2012 was 1,897 lbs./MWh (compared with the average rate for coal-fired plants of about 2,000 lbs./MWh). To develop the 2030 standard, EPA adjusted each state's baseline rate with four groups of actions, or "building blocks," that the agency determined states could reasonably achieve:

1) Increase efficiency of existing coal-fired steam units by 6%;
2) Replace some coal-fired generation with increased output from underutilized natural gas combined-cycle units;
3) Prevent projected losses of 6% of each state's existing nuclear generating capacity, complete five nuclear reactors currently under construction, and increase generation from renewable energy; and
4) Implement energy efficiency measures to reduce projected electricity demand.

The nuclear adjustment was made by calculating the projected annual electricity generation from reactors under construction within each state (if any), along with the electricity generated by 6% of each state's existing nuclear capacity, which EPA deems to be at risk of shutdown. These amounts are added to each state's total fossil fuel-fired electrical generation, resulting in a decrease in the $CO_2$ generation rate per megawatt-hour.[127]

In the Ohio example, the state has no reactors under construction but has two operating nuclear reactors, with total generating capacity of 2,150 megawatts. Six percent of that capacity is projected to generate 993,077 MWh per year, which EPA added to total generation from fossil fuel sources, renewable energy sources, and electricity generation avoided through efficiency measures, as described in the "building blocks." The increase in total electric generation spreads $CO_2$ emissions among more megawatt-hours, so the rate of $CO_2$/MWh goes down. The nuclear additions plus the other changes described above result in a proposed 2030 standard for Ohio of 1,338 lbs. of $CO_2$ per MWh—a 29.5% reduction from the unadjusted baseline.

Although the state standards for $CO_2$ emission rates would be binding, each state could meet them using whatever mix of options it chose. A state would not have to employ the same "building blocks" in its compliance plans that EPA used to calculate the standards. According to an EPA explanation, each state "is free to meet that goal in the way that works best for that state. It can rely more or less heavily on specific measures such as efficiency or renewable energy, or even pursue others such as increases in transmission efficiency or new gas generation."[128] Additional nuclear reactors beyond the five already under construction would also be an option.

Because of the state flexibility, the effect of the EPA proposal on nuclear power is inherently uncertain. EPA's emission rate methodology might encourage states to take steps to ensure that nuclear plants currently under construction were completed, and to prevent the shutdown of existing nuclear capacity, because if they did not, they would need to find alternative ways to meet the $CO_2$ emission rate standard. For example, if Ohio's two nuclear plants closed, the state would lose 993,077 MWh of zero-carbon electricity in its $CO_2$ rate formula and would need to increase other low-carbon electricity sources or take other steps beyond those already in the EPA building blocks.

The Nuclear Energy Institute, representing the nuclear industry, issued a statement praising the EPA proposal for recognizing "the need to maintain and expand the use of nuclear energy."[129] However, nuclear industry officials have expressed unhappiness that the nuclear "building block" includes only the 6% of existing nuclear capacity considered to be "at risk," rather than all existing nuclear capacity. According to a media report, the industry is concerned that including only 6% of existing nuclear in the state goal calculations would not provide states with enough incentive to prevent nuclear plants from being shut down and replaced by cheaper gas plants.[130] In the Ohio example, only 993,077 MWh of carbon-free generation would have to be replaced with other options if both the state's reactors were shut down, under the EPA proposal. But if all the state's nuclear electricity were included in the nuclear building block, then 16.95 million MWh (about 9% of the state's total 2012 generation) would have to be replaced if the Ohio reactors shut down (2,150 megawatts operating at 90% of capacity for a year).

On the other hand, a news release by the Nuclear Information and Resource Service, a group critical of the nuclear industry, contended that

EPA's proposed rule "would encourage states to provide ratepayer subsidies for continued operation of nuclear reactors that cannot compete economically in the current electricity marketplace."[131]

## Nuclear Power Research and Development

The Obama Administration's FY2015 funding request for nuclear energy research and development totaled $863.4 million. Including advanced reactors, fuel cycle technology, infrastructure support, and safeguards and security, the total nuclear energy request was $25.0 million (2.9%) below the FY2014 funding level. DOE's FY2015 nuclear R&D budget justification described the following major goals for the program:

- Improve the safety, reliability, and economics of nuclear power plants;
- Implement a "consent based" strategy for developing nuclear waste storage and disposal facilities;
- Develop improved waste management and fuel cycle technologies; and
- Understand and minimize the risks of nuclear proliferation and terrorism.

DOE's Office of Nuclear Energy would lead a major initiative announced in the FY2015 budget request to commercialize the Brayton cycle for commercial power plants. Called Supercritical Transformational Electric Power Generation (STEP), the initiative was proposed to be a joint effort by DOE's nuclear energy, fossil energy, and renewable energy programs. The Brayton cycle uses supercritical gas to drive electric generators rather than the steam cycle that dominates the industry today. DOE's budget justification predicted that Brayton-cycle power plants could reach efficiencies of up to 50%, compared with 33% for steam-cycle plants. The STEP program is to reach a 50-50 cost sharing agreement with the private sector in FY2015 to develop a 10 megawatt (electric) Brayton cycle pilot plant. Funding for the STEP initiative, provided within the nuclear energy budget, would total $27.5 million in FY2015.

The House Appropriations Committee had recommended $899.0 million for nuclear energy, $35.6 million above the Administration request. However, an amendment on the House floor (H.Amdt. 979) reduced the nuclear total by

$73.3 million to provide more funding for the Corps of Engineers. The amendment did not specify how the reduction would be allocated among Nuclear Energy programs.

The Senate subcommittee draft report recommended $777.0 million for nuclear energy programs, $86.4 million below the request. The draft rejected the Administration's $97 million request for small modular reactor licensing support and cut the $100.5 million request for reactor concepts R&D by 45%.

The House Appropriations Committee report agreed with the Administration's $27.5 million request for the STEP initiative and authorized DOE to "modify" the 50% cost-sharing goal for qualifying pilot plants. The draft Senate report also recommended the full STEP request but cautioned that the program should "be limited in scope, schedule, and cost."

## Reactor Concepts

The Reactor Concepts program area includes research on advanced reactors, including advanced small modular reactors, and research to enhance the "sustainability" of existing commercial light water reactors. The total FY2015 funding request for this program was $100.5 million, a reduction of $12.3 million from FY2014.

DOE proposed to combine the Small Modular Reactor (SMR) R&D and Advanced Reactor Concepts subprograms into the Advanced Reactor Technologies subprogram in FY2015. Funding for the combined subprogram would be $70.2 million, a reduction of $12.6 million from the combined subprograms in FY2014. Reactor concepts being developed by the Advanced Reactor Technology subprogram are generally classified as "Generation IV" reactors, as opposed to the existing fleet of commercial light water reactors, which are generally classified as generations II and III. Nuclear technology development under this program focuses on "fast reactors," using high-energy neutrons, fluoride salt-cooled high-temperature reactors, and high temperature gas-cooled reactors. International research collaboration in this area would continue under the Generation IV International Forum (GIF).

The House Appropriations Committee recommended $138.0 million for Reactor Concepts, $37.5 million more than the request and $25.0 million above FY2014. The committee agreed with the proposal to consolidate Advanced SMR research with Advanced Reactor Concepts. The House panel voted to boost funding for the Advanced Reactor Concepts subprogram to $101.0 million, adding funding for high temperature gas reactor research. As noted above, the House approved an amendment to reduce the Committee's recommended total nuclear funding level without specifying cuts in individual nuclear subprograms.

DOE's FY2015 request for the Light Water Reactor Sustainability subprogram was $30.3 million, $350,000 above the FY2014 appropriation. The House panel recommended $35.0 million. The program conducts research on extending the life of existing commercial light water reactors beyond 60 years, the maximum operating period currently licensed by the Nuclear Regulatory Commission (NRC). The program, which is cost-shared with the nuclear industry, studies the aging of reactor materials and analyzes safety margins of aging plants. This subprogram is also conducting research to understand the Fukushima disaster and to develop prevention and mitigation measures, according to the DOE justification.

The draft Senate bill and report would cut reactor concepts to $55.0 million, consisting of $49.2 million for the combined Advanced Reactor Technologies subprogram and $5.8 million for Light Water Reactor Sustainability. The draft report directed DOE to focus the Light Water Reactor Sustainability subprogram on "understanding of accident scenarios, such as those exhibited in the Fukushima Daiichi nuclear disaster."

## Small Modular Reactor Licensing Support

Rising cost estimates for large conventional nuclear reactors—widely projected to be $6 billion or more—have contributed to growing interest in proposals for small modular reactors (SMRs). Ranging from about 40 to 300 megawatts of electrical capacity, such reactors would be only a fraction of the size of current commercial reactors, which typically exceed 1,000 megawatts. Several modular reactors would be installed together to make up a power block with a single control room, under most concepts. Current SMR proposals would use a variety of technologies, including high-temperature gas technology and the light water (LWR) technology used by today's commercial reactors.

DOE requested $97.0 million for FY2015 to provide technical support for licensing small modular reactors, $13 million below the FY2014 funding level. Under the program, DOE is to pay up to half the costs associated with NRC design certification and licensing of selected SMRs, as well as for economic studies and other analyses that would support SMR deployment in general.

A consortium led by Babcock & Wilcox (B&W) was announced by DOE in November 2012 as the first award recipient under the program. However, B&W announced April 14, 2014, that it would reduce its spending on the project to $15 million per year and delay its design certification application to NRC indefinitely, citing a lack of investors and customer contracts.[132] Because of the project's slowdown, DOE reportedly stopped paying matching funds to B&W after the first quarter of 2014.[133]

DOE selected a second SMR to receive assistance under the program in December 2013. The NuScale Power SMR has a generating capacity of only 45 megawatts. Under the company's current concept, up to 12 reactors would be housed in a single pool of water, which would provide emergency cooling. The NuScale SMR is intended to be ready for commercial operation by around 2025, according to DOE.[134] The DOE budget justification contends that reduced funding for the SMR program will be sufficient for both the B&W and NuScale projects in FY2015.

Because of the uncertainty about the B&W SMR project, the House Appropriations Committee cut the SMR program to $54.5 million, with all of the remaining funding directed to the NuScale project. However, the committee said that it "will consider additional funding according to developments." The Senate draft report recommended no new funding for SMR licensing support, citing the availability of $85.0 million in prior-year funds that could be reprogrammed for the NuScale project.

Small modular reactors would go against the overall trend in nuclear power technology toward ever-larger reactors intended to spread construction costs over a greater output of electricity. Proponents of small reactors contend that they would be economically viable despite their far lower electrical output because modules could be assembled in factories and shipped to plant sites, with minimal on-site fabrication, and because their smaller size would allow for simpler and more effective safety systems. In addition, although modular plants might have similar or higher costs per kilowatt-hour than conventional large reactors, their ability to be constructed in smaller increments could reduce electric utilities' financial commitment and risk.

## Fuel Cycle Research and Development

The Fuel Cycle Research and Development Program conducts "long-term, science-based" research on a wide variety of technologies for improving the management of spent nuclear fuel, according to the DOE budget justification.

In general, the program is investigating ways to separate radioactive constituents of spent fuel for re-use or to be bonded into stable waste forms. The total FY2015 funding request for this program was $189.1 million, $2.9 million above the FY2014 appropriation.

The Administration requested a nearly one-third increase for the Used Nuclear Fuel R&D subprogram, from $60.0 million in FY2014 to $79.0 million in FY2015. This subprogram focuses on establishing a new spent fuel management system, consistent with the Administration's moves to terminate the previously authorized waste repository program at Yucca Mountain, NV. DOE released its Strategy for the Management and Disposal of Used Nuclear Fuel and High-Level Radioactive Waste in January 2013 that calls for a "consent-based siting process" for nuclear storage and disposal facilities. The Used Fuel subprogram would also conduct waste transportation analyses and research on potential waste repositories, including salt caverns and deep boreholes, according to the DOE justification. DOE also proposed that Congress provide mandatory appropriations for the spent fuel management program beginning in FY2018 to supplement discretionary appropriations. (See the "Nuclear Waste Management" section for more details.)

Other major research areas in the Fuel Cycle R&D Program include the development of accident-tolerant fuels for existing commercial reactors, evaluation of fuel cycle options, development of improved technologies to prevent diversion of nuclear materials for weapons, and technology to increase nuclear fuel resources, such as uranium extraction from seawater.

The House Appropriations Committee recommended $182.0 million for Fuel Cycle R&D, $7.1 million below the request and $4.5 million below FY2014.

The committee provided $60.1 million for accident-tolerant fuels research, $55.0 million for Used Nuclear Fuel Disposition, and $55.0 million for used-fuel disposition R&D, such as long-term dry cask storage.

The Senate draft report recommended $230.0 million for Fuel Cycle R&D, including $119.0 million for Used Nuclear Fuel Disposition. Within Used Fuel Disposition, $89.0 million would be provided for developing a consolidated spent fuel storage facility through a consent-based siting process. Such a storage facility could hold spent fuel from nuclear power plants until a permanent underground repository could be developed. The draft report also recommended $60.1 million for accident-tolerant fuels.

## NUCLEAR WASTE MANAGEMENT

One of the most controversial aspects of nuclear power is the disposal of radioactive waste, which can remain hazardous for thousands of years. Each nuclear reactor produces an annual average of about 20 metric tons of highly radioactive spent nuclear fuel, for a nationwide total of about 2,000 metric tons per year. U.S. reactors also generated an average of about 75,000 cubic meters of low-level radioactive waste per year during the past decade, including contaminated components and materials resulting from reactor decommissioning.[135]

The federal government is responsible for permanent disposal of commercial spent fuel (paid for with a fee on nuclear power production) and federally generated radioactive waste, while states have the authority to develop disposal facilities for most commercial low-level waste. Under the Nuclear Waste Policy Act (NWPA, 42 U.S.C. 10101, et seq.), spent fuel and other highly radioactive waste is to be isolated in a deep underground repository, consisting of a large network of tunnels carved from a geologic formation that has remained stable for hundreds of thousands of years. As amended in 1987, NWPA designated Yucca Mountain in Nevada as the only candidate site for the national repository. The act required DOE to begin taking waste from nuclear plant sites by 1998—a deadline that even under the most optimistic scenarios will be missed by more than 20 years. DOE filed a license application with NRC for the proposed Yucca Mountain repository in June 2008.

The Obama Administration "has determined that developing the Yucca Mountain repository is not a workable option and the Nation needs a different solution for nuclear waste disposal," according to the DOE FY2011 budget justification. To develop alternative waste management strategies, the Administration established the Blue Ribbon Commission on America's Nuclear Future, which issued its final report to the Secretary of Energy on January 26, 2012.[136] The Blue Ribbon Commission recommended that future efforts to develop nuclear waste facilities follow a "consent based" approach and be carried out by a new organization, rather than DOE. The Commission said the new nuclear waste entity should have "assured access" to the Nuclear Waste Fund, which holds fees collected from nuclear power plant operators to pay for waste disposal. Under NWPA, those funds cannot be spent without congressional appropriations.

DOE released its *Strategy for the Management and Disposal of Used Nuclear Fuel and High-Level Radioactive Waste* in January 2013 in response to the Blue Ribbon Commission report. The strategy calls for a pilot interim storage facility for spent fuel from closed nuclear reactors to open by 2021 and a larger storage facility, possibly at the same site, to open by 2025. A site for a permanent underground waste repository would be selected by 2026, and the repository would open by 2048. Storage and disposal sites would be selected by a new waste management organization through a consent-based process, as recommended by the Blue Ribbon Commission.[137]

DOE's Office of Nuclear Energy (NE) currently is responsible for civilian waste management activities. NE's Fuel Cycle R&D Program (discussed in the "Nuclear Power Research and Development" section above) includes funding under the Used Nuclear Fuel Disposition subprogram to begin implementing the DOE waste management strategy. DOE requested $79.0 million for the Used Fuel subprogram in FY2015, $19.0 million above the FY2012 funding level, and no funding for Yucca Mountain.

In approving the Energy and Water Development Appropriations bill for FY2014 (H.R. 2609), the House Appropriations Committee excoriated the Obama Administration's termination of the Yucca Mountain project as "blatant political maneuverings." The House-passed FY2015 energy and water bill included $150.0 million for DOE and $55.0 million for NRC to continue Yucca Mountain licensing activities. The House provided no funding to implement DOE's alternative nuclear waste management strategy.

The Senate Appropriations subcommittee recommended $119.0 million in FY2015 for Used Fuel and did not mention Yucca Mountain. The subcommittee's draft bill included a provision from the previous year that would authorize DOE to conduct a pilot program to develop one or more high level radioactive waste storage facilities, with the consent of state, local, and tribal governments.

Senator Wyden, along with Senators Murkowski, Feinstein, and Alexander, introduced legislation June 27, 2013, to redirect the nuclear waste program (S. 1240) along the lines recommended by the Blue Ribbon Commission. The bill would establish an independent Nuclear Waste Administration to develop nuclear waste storage and disposal facilities. Siting of such facilities would require the consent of the affected state, local, and tribal governments. The Nuclear Waste Administration could spend nuclear waste fees collected after the bill's enactment without the need for further appropriation. Fee collection would halt after 2025 if a waste facility had not been opened. The Energy and Natural Resources Committee held a hearing on the bill July 30, 2013.

DOE had filed a license application with NRC for the proposed Yucca Mountain repository in June 2008 but filed a motion to withdraw the application on March 3, 2010. An NRC licensing panel rejected DOE's withdrawal motion June 29, 2010, on the grounds that NWPA requires full consideration of the license application by NRC. The full NRC Commission deadlocked on the issue September 9, 2011, leaving the licensing panel's decision in place and prohibiting DOE from withdrawing the Yucca Mountain application. However, the commission ordered at the same time that the licensing process be suspended because of "budgetary limitations."[138] No funding was provided in FY2012 or FY2013 or requested for FY2014 or FY2015 to continue Yucca Mountain licensing activities. However, the U.S. Court of Appeals for the District of Columbia Circuit ruled on August 13, 2013, that NRC must continue work on the Yucca Mountain license application as long as funding is available. The Court determined that NRC had at least $11.1 million in previously appropriated funds for that purpose.[139]

NWPA required DOE to begin taking waste from nuclear plant sites by January 31, 1998. Nuclear utilities, upset over DOE's failure to meet that deadline, have won two federal court decisions upholding the department's obligation to meet the deadline and to compensate utilities for any resulting damages. Utilities have also won several cases in the U.S. Court of Federal Claims. DOE estimates that liability payments would eventually exceed $20 billion if DOE were to begin removing waste from reactor sites by 2020, the previous target for opening Yucca Mountain.[140]

Pursuant to a court ruling, DOE reportedly stopped collecting nuclear waste fees from nuclear power generators on May 16, 2014, eliminating about $750 million in revenues for the waste disposal program.[141] NWPA requires the Secretary of Energy to adjust the fees as necessary to cover the waste program's anticipated costs, but the U.S. Court of Appeals for the District of Columbia Circuit ruled that DOE's current waste plans are too vague to allow a reasonable estimate to be calculated and that DOE must therefore stop collecting the fee.[142]

NRC approved a final rule August 26, 2014, on continued storage of spent nuclear fuel.[143] The rule responds to a federal circuit court ruling on June 8, 2012, that struck down NRC's Waste Confidence Decision, which contained the agency's formal findings that waste generated by nuclear power plants will be disposed of safely. The court ruled that the Waste Confidence Decision required an environmental review under the National Environmental Policy Act and that NRC needed to consider the possibility that a permanent waste repository would never be built and to examine potential problems with waste storage pools.

The Waste Confidence Decision, first issued in 1984 and since updated twice, resulted from a 1979 federal circuit court ruling that required NRC to determine whether waste from nuclear facilities would be safely managed after their licenses expired. After the Waste Confidence Decision was struck down in 2012, NRC suspended final licensing decisions for new reactors, spent fuel storage facilities, and license renewals. Upon approving the continued storage rule to replace the Waste Confidence Decision, NRC ended its licensing suspension.[144]

## NUCLEAR WEAPONS PROLIFERATION

Renewed interest in nuclear power in much of the world has led to increased concern about nuclear weapons proliferation, because technology for making nuclear fuel can also be used to produce nuclear weapons material. Of particular concern are uranium enrichment, a process to separate and concentrate the fissile isotope uranium-235, and nuclear spent fuel reprocessing, which can produce weapons-useable plutonium.

The International Atomic Energy Agency (IAEA) conducts a safeguards program that is intended to prevent civilian nuclear fuel facilities from being used for weapons purposes, but not all potential weapons proliferators belong to the system, and there are ongoing questions about its effectiveness. Several proposals have been developed to guarantee nations without fuel cycle facilities a supply of nuclear fuel in exchange for commitments to forgo enrichment and reprocessing, which was one of the original goals of the Bush Administration's Global Nuclear Energy Partnership, now called the International Framework for Nuclear Energy Cooperation.[145]

Several situations have arisen throughout the world in which ostensibly commercial uranium enrichment and reprocessing technologies have been subverted for military purposes. In 2003 and 2004, it became evident that Pakistani nuclear scientist A.Q. Khan had sold sensitive technology and equipment related to uranium enrichment to states such as Libya, Iran, and North Korea. Although Pakistan's leaders maintain they did not acquiesce in or abet Khan's activities, Pakistan remains outside the Nuclear Nonproliferation Treaty (NPT) and the Nuclear Suppliers Group (NSG). Iran has been a direct recipient of Pakistani enrichment technology.

IAEA's Board of Governors found in 2005 that Iran's breach of its safeguards obligations constituted noncompliance with its safeguards agreement, and referred the case to the U.N. Security Council in February 2006. Despite repeated calls by the U.N. Security Council for Iran to halt enrichment and reprocessing-related activities, and imposition of sanctions, Iran continues to develop enrichment capability at Natanz and at a site near Qom disclosed in September 2009. Iran insists on its inalienable right to develop the peaceful uses of nuclear energy, pursuant to Article IV of the NPT. Interpretations of this right have varied over time. International talks with Iran over its nuclear program are ongoing. The case of Iran raises perhaps the most critical current question for strengthening the nuclear nonproliferation regime: How can access to sensitive fuel cycle activities (which could be used to produce fissile material for weapons) be circumscribed without further alienating non-nuclear weapon states in the NPT?

Leaders of the international nuclear nonproliferation regime have suggested ways of reining in the diffusion of such inherently dual-use technology, primarily through the creation of incentives not to enrich uranium or reprocess spent fuel. The international community is in the process of evaluating those proposals and may decide upon a mix of approaches. At the same time, there is debate on how to improve the IAEA safeguards system and its means of detecting diversion of nuclear material to a weapons program in the face of expanded nuclear power facilities worldwide.

## FEDERAL FUNDING FOR NUCLEAR ENERGY PROGRAMS

The following tables summarize current funding for DOE nuclear energy programs and NRC. The sources for the funding figures are Administration budget requests[146] and committee reports on the Energy and Water Development Appropriations Acts, which fund DOE and NRC. The House passed its version of the FY2015 Energy and Water bill (H.R. 4923, H.Rept. 113-486) on July 10, 2014. The Senate Subcommittee on Energy and Water Development Appropriations approved a draft FY2015 bill and report on July 24, 2014.[147] Pending enactment of full-year appropriations, FY2015 funding for these programs is being provided by a continuing resolution (P.L. 113-164).

**Table 2. Funding for the Nuclear Regulatory Commission (budget authority in millions of current dollars)**

| | **FY2012 Approp.** | **FY2013 Approp.**[a] | **FY2014 Approp.** | **FY2015 Request** | **FY2015 House** | **FY2015 Senate Subcom.** |
|---|---|---|---|---|---|---|
| Reactor Safetyb<br>Nuclear Materials and Waste | 800.1a<br>227.1 | —c<br>— | 811.4<br>232.5 | 815.2<br>232.2 | —c<br>— | —<br>— |
| Yucca Mountain | 0 | 0 | 0 | 0 | 55.0 | 0 |
| Licensing | | | | | | |
| Inspector General | 10.9 | 10.9 | 12.0 | 12.1 | 12.1 | 12.1 |
| **Total NRC budget authority** | 1,038.1 | 1,036.0 | 1,055.9 | 1,059.5 | 1,064.5 | 1,059.5 |
| —Offsetting fees | -909.5 | -909.5 | -930.7 | -935.2 | -890.3 | -935.2 |
| **Net appropriation** | **128.6** | **126.6** | **125.2** | **124.3** | **174.3** | **124.3** |

[a] FY2013 figures do not reflect March 1, 2013, sequester under P.L. 112-25.

[b] Subcategories from NRC budget request.

[c] Subcategories not specified.

**Table 3. DOE Funding for Nuclear Activities (Selected Programs) (budget authority in millions of current dollars)**

| | FY2012 Approp. | FY2013 Approp. | FY2014 Approp. | FY2015 Request | FY2015 House | FY2015 Senate Subcom. |
|---|---|---|---|---|---|---|
| Reactor Concepts | 115.5 | 104.8 | 113.0 | 100.5 | 138.0 | 55.0 |
| Small Modular Reactor | 67.0 | 62.7 | 110.0 | 97.0 | 54.5 | 0 |
| Licensing | | | | | | |
| Fuel Cycle R&D | 187.4 | 169.9 | 186.5 | 189.1 | 182.0 | 230.0 |
| Nuclear Energy Enabling | 74.9 | 67.9 | 71.1 | 78.2 | 101.0 | 73.5 |
| Technologies | | | | | | |
| International Nuclear | 3.0 | 2.8 | 2.5 | 3.0 | 3.0 | 3.0 |
| Energy Cooperation | | | | | | |
| Radiological Facilities | 69.9 | 65.4 | 25.0 | 5.0 | 5.0 | 25.0 |
| Management | | | | | | |
| Idaho Facilities | 155.0 | 145.0 | 196.6 | 185.9 | 206.0 | 185.9 |

| | FY2012 Approp. | FY2013 Approp. | FY2014 Approp. | FY2015 Request | FY2015 House | FY2015 Senate Subcom. |
|---|---|---|---|---|---|---|
| Management | | | | | | |
| Program Direction | 91.0 | 85.1 | 90.0 | 73.1 | 73.0 | 73.1 |
| Yucca Mountain repository | 0 | 0 | 0 | 0 | 150.0 | 0 |
| Total, Nuclear Energya | 765.4 | 708.4 | 889.2 | 863.4 | 826.7 | 777.0 |

[a] Excludes funding provided under other accounts.

[b] Nuclear energy total reduced by House floor amendment without specifying changes in subcategories.

# LEGISLATION IN THE 113TH CONGRESS

## H.R. 259 (Pompeo)/S. 2279 (Lee)

Energy Freedom and Economic Prosperity Act. Terminates nuclear energy production tax credit, among other provisions. House bill introduced January 15, 2013; referred to Committee on Ways and Means. Senate bill introduced May 1, 2014; referred to Committee on Finance.

## H.R. 1700 (Engel)

Nuclear Disaster Preparedness Act. Requires the President to issue guidance for federal response to nuclear disasters, covering specific topics listed in the bill. Introduced April 24, 2013; referred to Committee on Transportation and Infrastructure.

## H.R. 2081 (Thornberry)

No More Excuses Energy Act of 2013. Includes provisions to prohibit NRC from considering nuclear waste storage when licensing new nuclear facilities, and to establish a tax credit for obtaining nuclear component manufacturing certification. Introduced May 21, 2013; referred to multiple committees.

## H.R. 2609 (Frelinghuysen)/S. 1245 (Feinstein)

Energy and Water Development and Related Agencies Appropriations Act, 2014. Provides funding for DOE nuclear programs and NRC. House bill introduced July 2, 2013; reported as original measure by Committee on Appropriations July 2, 2013 (H.Rept. 113-135); passed House July 10, 2013, by vote of 227-198. Senate bill introduced June 27, 2013; reported as original measure by Committee on Appropriations June 27, 2013 (S.Rept. 113-47). FY2014 Energy and Water Development funding provided by Division D of P.L. 113-76, Consolidated Appropriations Act, 2014.

## H.R. 2712 (Lowey)

Nuclear Power Licensing Reform Act of 2013. Requires evacuation planning within 50 miles of U.S. nuclear power plants and that reactor license renewals be subject to the same standards that would apply to new reactors. Introduced July 17, 2013; referred the Committee on Energy and Commerce.

## H.R. 2861 (Lowey)

Requires NRC to distribute safety-related fines collected from nuclear facilities to the counties in which the facilities are located to maintain radiological emergency preparedness plans. Introduced July 30, 2013; referred to Committee on Energy and Commerce.

## H.R. 3354 (Engel)

Dry Cask Storage Act. Requires spent fuel at nuclear power plants to be moved from spent fuel pools to dry casks after it has sufficiently cooled. Costs of the fuel transfers would be offset by a reduction in nuclear waste fees owed to the federal government. Introduced October 28, 2013; referred to Committee on Energy and Commerce.

### H.R. 3766 (Ros-Lehtinen)

Requires congressional approval of U.S. peaceful nuclear cooperation agreements with countries that do not agree to forgo enrichment and reprocessing. Introduced December 12, 2013; referred to Committees on Foreign Affairs and Rules.

### H.R. 3895 (Duncan)

Energy Exploration and Production to Achieve National Demand (EXPAND) Act. Among other provisions, repeals nuclear production tax credit, authorizes expedited nuclear licensing procedures, requires NRC to develop "technology neutral" guidelines for licensing advanced nuclear plants, requires an accelerated schedule for the Next Generation Nuclear Plant, and limits fees for uranium mining on federal lands. Introduced January 16, 2014; referred to multiple committees.

### H.R. 4522 (Van Hollen)/S. 2271 (Murphy)

Establishes a Green Bank to finance clean energy, including nuclear projects. Both bills introduced April 30, 2014; House bill referred to Committee on Ways and Means and to Energy and Commerce for specific provisions; Senate bill referred to Committee on Finance.

### H.R. 4869 (Lummis)

Department of Energy Research and Development Act of 2014. Authorizes DOE research and development programs, including nuclear energy. Introduced June 13, 2014; referred to Committee on Science, Space, and Technology.

### H.R. 4923 (Simpson)

Energy and Water Development and Related Agencies Appropriations Act, 2015. Provides appropriations for DOE nuclear programs and NRC.

Introduced June 20, 2014; reported as an original measure June 20, 2014, by the Committee on Appropriations (H.Rept. 113-486). Passed House July 10, 2014, by vote of 253-170.

### H.R. 4956 (Walz)

American Energy Opportunity Act of 2014. Among other provisions, establishes the Carbon Free Technology and Nuclear Energy Reserve that could be used to offset the cost of loan guarantees for commercial nuclear power plants, the disposition and recycling or reprocessing of spent fuel from nuclear power plants, and the financing of long-term safe storage of spent fuel. Introduced June 24, 2014; referred to multiple committees.

### H.R. 5322 (Blackburn)

Consolidate Heavy-handed and Outdated Programs Act of 2014. Consolidates nuclear energy programs under a new Department of Energy and the Environment. Introduced July 31, 2014; referred to multiple committees.

### S. 1240 (Wyden)

Nuclear Waste Administration Act of 2013. Establishes an independent Nuclear Waste Administration to develop nuclear waste storage and disposal facilities. Siting of such facilities would require the consent of the affected state, local, and tribal governments. The Nuclear Waste Administration could spend nuclear waste fees collected after the bill's enactment without the need for further appropriation. Fee collection would halt after 2025 if a waste facility had not been opened. Introduced June 27, 2013; referred to Committee on Energy and Natural Resources. Full committee hearing held July 30, 2013.

### S. 1519 (Vitter)

Nuclear Regulatory Commission Reorganization Plan Codification and Complements Act. Specifies functions and authorities of the Chairman and Commissioners of NRC. Specifies that any commissioner may request a vote on whether a particular issue should be reserved for the Chairman or handled by the full Commission. Introduced September 18, 2013; referred to Committee on Environment and Public Works.

## S. 2324 (Boxer)

Safe and Secure Decommissioning Act of 2014. Prohibits NRC from waiving emergency preparedness and security requirements at nuclear reactors that are undergoing decommissioning until all of such a reactor's spent nuclear fuel has been transferred from storage pools into dry casks. Introduced May 13, 2014; referred to Committee on Environment and Public Works.

## S. 2325 (Markey)

Dry Cask Storage Act of 2104. Requires each nuclear reactor to submit a plan to NRC for tranferring spent fuel currently stored in pools to dry casks within seven years. After the seven year period, additional spent fuel must be transferred to dry casks within a year after it has been determined to be sufficiently cool. Emergency planning zones must be expanded from 10 to 50 miles in radius around any reactor that is determined by NRC to be out of compliance with its spent fuel transfer plan. Introduced May 13, 2014; referred to Committee on Environment and Public Works.

## S. 2326 (Sanders)/H.R. 4667 (Welch)

Nuclear Plant Decommissioning Act of 2014. Provides for consultation with state and local governments, and the consideration of state and local concerns, during NRC approval of post-shutdown decommissioning activities reports, and requires reactors undergoing decommissioning to comply with state laws and standards. Senate bill introduced May 13, 2014; referred to Committee on Environment and Public Works. House bill introduced May 15, 2014; referred to Committee on Energy and Commerce.

## End Notes

[1] Mirshak, Meg, "Energy Secretary Moniz Visits Vogtle to Finalize Loan Guarantee," *Augusta Chronicle*, February 21, 2014, http://chronicle.augusta.com/news/metro/2014-02-20/energy-secretary-moniz-visits-vogtle-finalize-loanguarantee.

[2] Southern Company, "Southern Company Subsidiary, DOE Finalize Vogtle Nuclear Project Loan Guarantees," news release, February 20, 2014, http://www.prnewswire.com/news-releases/southern-company-subsidiary-doe-finalizevogtle-nuclear-project-loan-guarantees-246395221.html.

[3] See CRS Insight.

[4] Senate Committee on Appropriations, "FY15 Subcommittee Reported Bill and Draft Report," July 24, 2014, http://www.appropriations.senate.gov/news/fy-2015-ew-subcommittee-reported-bill-and-draft-report.

[5] Nuclear Regulatory Commission, "NRC Certifies GE-Hitachi New Reactor Design," news release, September 16, 2014, http://www.nrc.gov/reading-rm/doc-collections/news/2014/.

[6] Environmental Protection Agency, "Clean Power Plan Proposed Rule," http://www2.epa.gov /carbon-pollutionstandards/clean-power-plan-proposed-rule.

[7] DOE, *Strategy for the Management and Disposal of Used Nuclear Fuel and High-Level Waste*, January 2013, http://energy.gov/sites/prod/files/ Strategy%20for%20the%20Management %20and%20Disposal%20of%20Used%20Nuclear%20Fuel%20and%20High %20Level%20Radioactive%20Waste.pdf.

[8] U.S. Court of Appeals for the District of Columbia Circuit, In re: Aiken County et al., No. 11-1271, writ of mandamus, August 13, 2013, http://www.cadc.uscourts.gov /internet /opinions.nsf/BAE0CF34F762EBD985257BC6004DEB18/$file/11-1271-1451347.pdf.

[9] Hiruo, Elaine, "DOE Implements Court-Ordered Suspension of Nuclear Waste Fee," *NuclearFuel*, May 26, 2014.

[10] U.S. Court of Appeals for the District of Columbia Circuit, National Association of Regulatory Utility Commissioners v. U.S. Department of Energy, No. 11-1066, decided November 19, 2013, http://www.cadc.uscourts.gov/internet/opinions.nsf/2708C01ECFE3109 F85257C 280053406E/$file/11-1066- 1466796.pdf.

[11] NRC, "Waste Confidence—Continued Storage of Spent Nuclear Fuel," proposed rule, 78 *Federal Register* 56776, September 13, 2013.

[12] NRC, "NRC Approves Final Rule on Spent Fuel Storage and Ends Suspension of Final Licensing Actions for Nuclear Plants and Renewals," news release, August 26, 2014, http://pbadupws.nrc.gov/docs/ML1423/ ML14238A326.pdf.

[13] Oral Testimony of Energy Secretary Steven Chu at the House Energy and Commerce Committee – As Prepared for Delivery, March 16, 2011, http://www.energy.gov /news/10178.htm.

[14] Energy Information Administration, *Electricity Data Browser*, "Net Generation for All Sectors, Annual," viewed September 4, 2013, http://www.eia.gov/electricity/data/browser.

[15] U.S. Nuclear Regulatory Commission, *Information Digest 2023-2014*, NUREG-1350, Vol. 25, August 20013, p. 32, http://www.nrc.gov/reading-rm/doc-collections/nuregs/staff/sr1350 /v25/sr1350v25.pdf.

[16] Mary Powers, "Credit Agencies See Watts Bar-2 Cost Impact," *Nucleonics Week*, April 12, 2012, p. 1.

[17] Nuclear Regulatory Commission, "In the Matter of Tennessee Valley Authority (Bellefonte Nuclear Plant Units 1 and 2)," 74 *Federal Register* 10969, March 13, 2009.

[18] Tennessee Valley Authority, "TVA Board Implements Vision," press release, August 18, 2011, http://www.tva.com/ news/releases/julsep11/board_meeting/index.htm.

[19] Tennessee Valley Authority, "TVA Announces Budget Reduction for Bellefonte Plant," press release, June 12, 2013, http://www.tva.com/news/releases/aprjun13/bellefonte_budget.html.

[20] Energy Information Administration, *Electric Power Monthly*, Net Generation by Energy Source, February 2014, http://www.eia.gov/electricity/monthly/epm_table_grapher.cfm?t=epmt_1_01. Net generation excludes electricity used for power plant operation.

[21] Nuclear Regulatory Commission, *Information Digest, 2013–2014*, op. cit.

[22] EIA, *Electricity Data Browser,* op. cit.

[23] All of today's 100 operating U.S. commercial reactors were ordered from 1963 through 1973; see "Historical Profile of U.S. Nuclear Power Development," U.S. Council for Energy Awareness, 1992.

[24] For a comparison of generating costs, see CRS Report RL34746, *Power Plants: Characteristics and Costs*, by Stan Mark Kaplan.

[25] Energy Information Administration, *Electric Power Monthly*, "Capacity Factors for Utility-Scale Generators Not Primarily Using Fossil Fuels," February 2014, http://www.eia.gov/electricity/monthly/epm_table_grapher.cfm?t= epmt_6_07_b.

[26] Nuclear Regulatory Commission, "Status of License Renewal Applications and Industry Activities," May 5, 2014, http://www.nrc.gov/reactors/operating/licensing/renewal /applications.html.

[27] Nuclear Regulatory Commission, New Reactors, http://www.nrc.gov/reactors/new-reactors.html.

[28] Southern Company, "Smart Power," http://www.southerncompany.com/smart_energy/smart_ power_vogtlekemper.html.

[29] South Carolina Electric and Gas Company, "V.C. Summer Nuclear Station Units 2 and 3, Quarterly Report to the South Carolina Office of Regulatory Staff," June 30, 2013, http://www.scana.com/NR/rdonlyres/FD5FC097-3956-48A6-9098-2C2D115C8512/0/NND QuarterlyReport2013Q2FINAL81413PUBLIC.pdf.

[30] Carr, Housely, "SCE&G Talks with Contractors on Summer Will Be 'Tense,' Analyst Says," *Nucleonics Week*, August 14, 2014, p. 5.

[31] Energy Information Administration, "Updated Capital Cost Estimates for Electricity Generation Plants," November 2010, http://www.eia.gov/oiaf/beck_plantcosts/index.html.

[32] South Carolina Electric & Gas Company, "Petitions for Updates and Revisions to the Capital Cost Schedule and the Construction Schedule," before the Public Service Commission of South Carolina, February 29, 2012, http://www.scana.com/NR/rdonlyres/35AAED95-5226-416A-8DC2-0743BC93B911/0/ 2012PetitiontoUpdateCostSchedules.pdf. Total cost based on SCE&G ownership of 55%.

[33] Southern Company, "Smart Power," http://www.southerncompany.com/smart_energy /smart_power_vogtlekemper.html. Total cost based on Southern Company's 45.7% ownership.

[34] Duke Energy, "Duke Energy Reaches Revised Multi-year Settlement with Florida Consumer Advocates," news release, August 1, 2013, http://www.duke-energy.com/news/releases /2013080101.asp.

[35] NRG Energy, "NRG Energy, Inc. Provides Greater Clarity on the South Texas Nuclear Development Project," press release, April 19, 2011, http://phx.corporate-ir.net/External.File?item=UGFyZW50SUQ9OTAwMzB8Q2hpbGRJRD0tMXxUeXBl PTM=&t=1.

[36] Nuclear Regulatory Commission, "NRC Certifies GE-Hitachi New Reactor Design," news release, September 16, 2014, http://www.nrc.gov/reading-rm/doc-collections/news/2014/.

[37] Jeff Beattie, "Southeast Utilities Seek Partners to Hedge Nuclear Bets," *Energy Daily*, October 5, 2010, p. 1.

[38] Tennessee Valley Authority, "Single Nuclear Unit at the Bellefonte Plant Site," fact sheet, http://www.tva.gov/ environment/reports/blnp/index.htm.

[39] Constellation Energy, "Constellation Energy Releases Statement Regarding U.S. Department of Energy Loan Guarantee," press release, October 9, 2010, http://ir.constellation.com /releasedetail.cfm?ReleaseID=516614.

[40] Letter from Michael J. Wallace, Vice Chairman and Chief Operating Officer, Constellation Energy, to Thomas Piquemal, Group Executive Vice President, Finance, EDF, October 15, 2010, http://files.shareholder.com/downloads/ CEG/1036755503x0x410084/e27369a0-ce85-432f-bfad-e17ddce4f8f2/101510_-_EDF_letter.pdf; Unistar, "EDF and Constellation Energy Announce Comprehensive Agreement," press release, October 27, 2010, http://press.edf.com/ press-releases/all-press-releases/2010/edf-and-constellation-energy-announce-comprehensive-agreement-82018.html& return=42873.

[41] NRC, "New Reactor Licensing Applications," August 15, 2014, http://www.nrc.gov/reactors /new-reactors/newlicensing-files/new-rx-licensing-app-legend.pdf.

[42] World Nuclear Association, "Fukushima Accident 2011," September 2014, http://www.world-nuclear.org/info/Safety-and-Security/Safety-of-Plants/Fukushima-Accident-2011. Radioactivity units converted from 940 petabecquerels of Iodine-131 equivalent.

[43] World Nuclear Association, "Chernobyl Accident 1986," June 2013, http://www.world-nuclear.org/info/Safety-andSecurity/Safety-of-Plants/Chernobyl-Accident.

[44] Nuclear Regulatory Commission, "Backgrounder on the Three Mile Island Accident," April 25, 2014, http://www.nrc.gov/reading-rm/doc-collections/fact-sheets/3mile-isle.html.

[45] French Institut de Radioprotection et de Surete Nucleaire (IRSN), Assessment on the 66th Day of Projected External Doses for Populations Living in the North-West Fallout Zone of the Fukushima Nuclear Accident, Report DRPH/2011- 10, p. 27, http://hps.org/documents /irsn_fukushima_report.pdf.

[46] Organisation for Economic Co-operation and Development, Nuclear Energy Agency, *2013 NEA Annual Report*, p. 8. CRS calculations based on data in Table 3.

[47] NRC, "Radiation Protection and the NRC," http://www.nrc.gov/reading-rm/doc-collections/nuregs/brochures/ br0322/#intro.

[48] See also Institute of Nuclear Power Operations, *Special Report on the Nuclear Accident at the Fukushima Daiichi Nuclear Power Station*, INPO 11-005, November 2011, available from the Nuclear Energy Institute at http://www.nei.org/resourcesandstats/document library/safetyandsecurity/reports/special-report-on-the-nuclearaccident-at-the-fukushima-daiichi-nuclear-power-station.

[49] Nuclear Regulatory Commission, "Nuclear Regulatory Commission Directs Staff on Continuing Agency Response to Japan Events; Adjust Commission Schedule," press release, March 23, 2011, http://pbadupws.nrc.gov/docs/ML1108/ ML110821123.pdf.

[50] Near-Term Task Force Review of Insights from the Fukushima Dai-ichi Accident, *Recommendations for Enhancing Reactor Safety in the 21st Century*, Nuclear Regulatory Commission, Washington, DC, July 12, 2011, http://pbadupws.nrc.gov/docs/ML1118 /ML111861807.pdf.

[51] NRC, "Recommended Actions to Be Taken Without Delay from the Near-Term Task Force Report," SECY-11- 0124, September 9, 2011.

[52] NRC, "Staff Requirements – SECY-11-0124 – Recommended Actions to Be Taken Without Delay from the Near-Term Task Force Report," October 18, 2011, http://pbadupws.nrc.gov/docs/ML1126/ML11269A204.pdf.

[53] NRC, "What Are the Lessons Learned from Fukushima?," September 10, 2014, http://www.nrc.gov/reactors/ operating/ops-experience/japan-dashboard/priorities.html.

[54] R.W. Borchardt, NRC Executive Director for Operations, "Prioritization of Recommended Actions to Be Taken in Response to Fukushima Lessons Learned," SECY-11-0137, October 3, 2011.

[55] 10 CFR 50.47, Emergency Plans.

[56] Nuclear Regulatory Commission, "Emergency Preparedness & Response," website, http://www.nrc.gov/about-nrc/ emerg-preparedness.html.

[57] Japanese Ministry of Education, Culture, Sports, Science, and Technology (MEXT), "Readings of Integrated Dose at Monitoring Post out of 20 Km Zone of Fukushima Dai-ichi NPP," data series, http://www.mext.go.jp/english/incident/ 1304275.htm.

[58] Nuclear Regulatory Commission, "NRC Provides Protective Action Recommendations Based on U.S. Guidelines," press release, March 16, 2011, http://pbadupws.nrc.gov/docs /ML1108/ML110800133.pdf.

[59] Nuclear Regulatory Commission, "Enhancements to Emergency Planning Regulations," Final rule, *Federal Register*, November 23, 2011, p. 72560.

[60] Nuclear Regulatory Commission, "Final Rule: Enhancements to Emergency Preparedness Regulations," SECY-11- 0053, April 8, 2011, http://www.nrc.gov/reading-rm/doc-collections/commission/secys/2011/2011-0053scy.pdf.

[61] Nuclear Energy Institute, "Myths & Facts About Safety," viewed September 16, 2014, .http://www.nei.org/ Knowledge-Center/Backgrounders/Fact-Sheets/Myths-Facts-About-Safety.

[62] Nuclear Regulatory Commission, "Backgrounder on the Three Mile Island Accident," April 25, 2014, http://www.nrc.gov/reading-rm/doc-collections/fact-sheets/3mile-isle.html.

[63] Evelyn O. Talbott et al., "Long Term Follow-Up of the Residents of the Three Mile Island Accident Area: 1979- 1998," Environmental Health Perspectives, March 2003, pp. 341-348, http://www.ncbi.nlm.nih.gov/pmc/articles/ PMC1241392.

[64] National Research Council, Committee to Assess the Health Risks from Exposure to Low Levels of Ionizing Radiation, *Beir VII: Health Risks from Exposure to Low Levels of Ionizing Radiation*, 2006. Available at http://www.nap.edu.

[65] National Academy of Sciences, Nuclear and Radiation Studies Board, "Analysis of Cancer Risks in Populations near Nuclear Facilities, Phase 2: Pilot Planning," http://nas-sites.org/cancerriskstudy/.

[66] NRC, "Safety Goals for the Operations of Nuclear Power Plants," policy statement, *Federal Register*, August 21, 1986, p. 30028, http://www.nrc.gov/reading-rm/doc-collections/commission/policy/51fr30028.pdf.

[67] NRC Staff Requirements Memorandum on SECY-89-102, "Implementation of the Safety Goals," Memorandum to James M. Taylor from Samuel J. Chilk, June 15, 1990, http://pbadupws.nrc.gov/docs/ML0037/ML003707881.pdf.

[68] U.S. NRC, Regulatory Guide 1.174, "An Approach for Using Probabilistic Risk Assessment in Risk-Informed Decisions on Plant-Specific Changes to the Licensing Basis," Revision 1, November 2002, http://www.nrc.gov/ reading-rm/doc-collections/reg-guides/power-reactors/rg/01-174.

[69] Public Citizen Energy Program, "The Myth of Nuclear Safety," http://www.citizen.org/cmep /energy_enviro_nuclear/ nuclear_power_plants/reactor_safety/articles.cfm?ID=4454.

[70] The Chernobyl Forum: 2003-2005, *Chernobyl's Legacy: Health, Environmental and Socio-Economic Impacts,* International Atomic Energy Agency, April 2006.

[71] Greenpeace. *The Chernobyl Catastrophe: Consequences on Human Health*, April 2006, p. 10.

[72] Nuclear Energy Institute, *Legislative Proposal to Help Meet Climate Change Goals by Expanding U.S. Nuclear Energy Production*, Washington, DC, October 28, 2009, p. 5, http://www.nei.org/resourcesandstats/documentlibrary/ newplants/policybrief/2009-nuclear-policy-initiative.

[73] For more information about the NRC reactor oversight process, see http://www.nrc.gov/NRR /OVERSIGHT/ ASSESS/index.html.

[74] NRC Office of Public Affairs, *NRC Approves Final Rule Amending Security Requirements*, News Release No. 07- 012, January 29, 2007.

[75] Nuclear Regulatory Commission, *Final Rule—Consideration of Aircraft Impacts for New Nuclear Power Reactors, Commission Voting Record*, SECY-08-0152, February 17, 2009.

[76] Nuclear Regulatory Commission, "Consideration of Aircraft Impacts for New Nuclear Power Reactors," Final Rule, 74 *Federal Register* 28111, June 12, 2009. This provision is codified at 10 CFR 50.150.

[77] Nuclear Regulatory Commission, "NRC Approves Final Rule Expanding Security Requirements for Nuclear Power Plants," press release, December 17, 2008, http://www.nrc.gov/reading-rm/doc-collections/news/2008/08-227.html.

[78] Nuclear Regulatory Commission, "Power Reactor Security Requirements," Final Rule, 74 *Federal Register* 13925, March 27, 2009.

[79] Sharp, David, "NRC Signs Off on Maine Yankee's Decommissioning," *Associated Press*, October 3, 2005.

[80] E-mail communication from Bob Capstick, Connecticut Yankee Atomic Power Company, August 28, 2008.

[81] Nuclear Regulatory Commission, "NRC Releases Rancho Seco Nuclear Plant for Unconditional Use," press release, October 7, 2009, http://www.nrc.gov/reading-rm/doc-collections/news/2009/09-165.html.

[82] "20 Years Later, Rancho Seco Ready for Final Shutdown," *Sacramento County Herald*, June 9, 2009.

[83] Edison International, "SCE Submits Nuclear Plant Decommissioning Plan to Federal Regulator," news release, September 23, 2014, http://newsroom.edison.com/releases/sce-submits-nuclear-plant-decommissioning-plan-to-federalregulator.

[84] NRC, *Information Digest, 2013-2014*, NUREG-1350, Volume 25, Appendix C, http://www.nrc.gov/reading-rm/doccollections/nuregs/staff/sr1350.

[85] Gutherman Technical Service, "2011 Used Fuel Data," January 14, 2012.

[86] DOE Office of Civilian Radioactive Waste Management, *Report to Congress on the Demonstration of the Interim Storage of Spent Nuclear Fuel from Decommissioned Nuclear Power Reactor Sites*, DOE/RW-0596, Washington, DC, December 2008, p. 1, http://www.energy.gov/media/ES_Interim_Storage_Report_120108.pdf.

[87] American Nuclear Insurers, "Need for Nuclear Liability Insurance," January 2010, http://www.nuclearinsurance.com/library/Nuclear%20Liability%20in%20the%20US.pdf.

[88] NRC, "Inflation Adjustments to the Price-Anderson Financial Protection Regulations," 79 *Federal Register* 41835, July 12, 2013.

[89] As of June 2014. See NRC, "Backgrounder on Nuclear Insurance and Disaster Relief," http://www.nrc.gov/readingrm/doc-collections/fact-sheets/nuclear-insurance.html. NRC can exempt shutdown reactors from the retrospective premiums after their spent fuel has sufficiently cooled. See NRC, "Exemption Requests to Reduce Liability Insurance

Coverage for Decommissioning Reactors after Transfer of All Spent Fuel from a Spent Fuel Pool to Dry Cask Storage, SECY-04-0176, September 29, 2004, http://www.nrc.gov/reading-rm/doc-collections/commission/secys/2004/.

[90] Reactors smaller than 100 megawatts must purchase an amount of liability coverage determined by NRC but are not subject to retrospective premiums. Total liability for those reactors is limited to $560 million, with the federal government indemnifying reactor operators for the difference between that amount and their liability coverage (Atomic Energy Act Sec. 170 b. and c.).

[91] Department of Energy, "Adjusted Indemnification Amount," 74 *Federal Register* 52793, October 14, 2009.

[92] Ellen Vancko, Union of Concerned Scientists, "The Impact of Fukushima on the US Nuclear Power Industry," presentation to the Center for Strategic and International Studies Conference on Nuclear Safety and Fukushima, April 7, 2011, https://csis.org/files /attachments/110407_vancko_nuclear_safety_0.pdf.

[93] International Atomic Energy Agency, Latest Status of Convention on Supplementary compensation for Nuclear Damage, July 7, 2014, https://www.iaea.org/Publications/Documents/Infcircs/1998/infcirc567.shtml.

[94] Ostroff, Jim, "Nuclear Compensation Convention on the Verge of Taking Effect," *Nucleonics Week*, December 12, 2013, p. 1.

[95] Nuclear Energy Institute, "NEI Unveils Package of Policy Initiatives Needed to Achieve Climate Change Goals," press release, October 26, 2009, http://www.nei.org/newsandevents /newsreleases/nei-unveils-package-of-policyinitiatives-needed-to-achieve-climate-change-goals/.

[96] Nuclear Information and Resource Service, "Senate Appropriators Lard President Obama's Stimulus Package with up to $50 Billion in Nuclear Reactor Pork," press release, January 30, 2009, http://www.nirs.org/press/01-30-2009/1.

[97] Department of the Treasury, Internal Revenue Service, *Internal Revenue Bulletin*, No. 2006-18, "Credit for Production From Advanced Nuclear Facilities," Notice 2006-40, May 1, 2006, p. 855.

[98] Energy Information Administration, *Status of Potential New Commercial Nuclear Reactors in the United States*, February 19, 2009.

[99] Nuclear Energy Institute, *Legislative Proposal to Help Meet Climate Change Goals by Expanding U.S. Nuclear Energy Production*, Washington, DC, October 28, 2009, p. 4, http://www.nei.org/resourcesandstats/documentlibrary/ newplants/policybrief/2009-nuclear-policy-initiative.

[100] Department of Energy, "Standby Support for Certain Nuclear Plant Delays," *Federal Register*, August 11, 2006, p. 46306.

[101] Freebairn, William, "Nuclear Tax Credits, Previously Low-Profile, Might Draw Scrutiny: Lobbysists," *Nucleonics Week*, May 17, 2012.

[102] Thomas B. Cochran and Christopher E. Paine, *Statement on Nuclear Developments Before the Committee on Energy and Natural Resources, United States Senate*, Natural Resources Defense Council, March 18, 2009, http://energy.senate.gov/public/index.cfm?Fuse Action=Hearings.Testimony&Hearing_ID=f25ddd10-c1f5-9e2e-528ec4321cca4c1b& Witness_ID=9f14a78d-58d0-43fb-bf5b-21426d1d888e.

[103] The FY2009 omnibus funding act (P.L. 111-8) increased DOE's total loan guarantee authority for specified technology categories to $47 billion, in addition to $4 billion in general authority provided in FY2007. Of the $47 billion, $18.5 billion continued to be reserved for nuclear power as provided in earlier appropriations, and $2 billion was for uranium

enrichment. Previous time limits on the loan guarantee authority were eliminated. DOE later allocated an additional $2 billion to enrichment projects from the $4 billion in general authority.

[104] Southern Company, "Federal Loan Guarantees Mark Latest Milestone in Historic Vogtle Project," news release, February 20, 2014.

[105] Congressional Budget Office, *S. 1321, Energy Savings Act of 2007*, CBO Cost Estimate, Washington, DC, June 11, 2007, pp. 7-9, http://www.cbo.gov/ftpdocs /82xx/doc8 206/s1321.pdf.

[106] Nuclear Energy Institute, "Loan Guarantees for the Vogtle Nuclear Power Project—Understanding the Credit Subsidy Cost," 2014, http://www.nei.org/Master-Document-Folder/Backgrounders/Fact-Sheets/Loan-Guarantees-forthe-Vogtle-Nuclear-Power-Proje?feed=factsheet.

[107] Richard Caperton, *Protecting Taxpayers from a Financial Meltdown*, Center for American Progress, Washington, DC, March 8, 2010, p. 2, http://www.americanprogress.org /issues/green/report/2010/03/08/7400/protecting-taxpayersfrom-a-financial-meltdown/.

[108] Letter from Michael J. Wallace, Vice Chairman and Chief Operating Officer, Constellation Energy, to Dan Poneman, Deputy Secretary of Energy, October 8, 2010, http://media.washingtonpost.com/wp-srv/hp/ssi/wpc/constellationenergy.PDF?sid=ST 2010100900005.

[109] http://www.lgprogram.energy.gov/keydocs.html.

[110] Marvin S. Fertel, *Statement for the Record to the Committee on Energy and Natural Resources, U.S. Senate*, Nuclear Energy Institute, March 18, 2009, p. 9, http://energy.senate.gov/public/index.cfm?FuseAction= Hearings.Testimony&Hearing_ID=f25ddd10-c1f5-9e2e-528e-c4321cca4c1b&Witness_ID=4de5e2df-53fe-49ba-906e9b69d3674e41.

[111] Department of Energy, "800 to 1000 New Jobs Coming to Piketon," press release, July 28, 2009, http://www.lgprogram.energy.gov/press/072809.pdf.

[112] Maureen Conley, "DOE Finds $2 Billion More for SWU Plant Loan Guarantees," *NuclearFuel*, April 5, 2010, p. 3.

[113] Department of Energy, "DOE Offers Conditional Loan Guarantee for Front End Nuclear Facility in Idaho," press release, May 20, 2010, http://www.energy.gov/news/8996.htm.

[114] Maureen Conley, "Congress 'Frustrated' by Inaction on USEC Loan Guarantee," *NuclearFuel*, October 31, 2011, p. 8.

[115] All FY3013 figures are pre-sequester.

[116] USEC Inc., "Five-Party Arrangement Extends Paducah Gaseous Diffusion Plant Enrichment Operations," press release, May 15, 2012, http://www.usec.com/news/five-party-arrangement-extends-paducah-gaseous-diffusion-plantenrichment-operations. The depleted uranium consists of "high assay" tails, which have relatively high levels of fissile U-235.

[117] USEC Inc., "Funding," web page, http://www.usec.com/american-centrifuge/what-american-centrifuge/plant/ funding.

[118] Department of Energy, "Obama Administration Announces Major Step Forward for the American Centrifuge Plant," press release, June 13, 2012, http://energy.gov/articles/obama-administration-announces-major-step-forwardamerican-centrifuge-plant.

[119] USEC, "Court Confirms USEC Inc. Plan of Reorganization," news release, September 5, 2014, http://www.usec.com/news/court-confirms-usec-inc-plan-reorganization.

[120] Hiruo, Elaine, "Low Volume, Non-Production Costs Driving Net Loss, USEC Says," *NuclearFuel*, August 18, 2014.

[121] DOE, "Department of Energy Issues Draft Loan Guarantee Solicitation for Advanced Nuclear Energy Projects," news release, September 30, 2014, http://www.energy.gov/articles /department-energy-issues-draft-loan-guaranteesolicitation-advanced-nuclear-energy-projects.

[122] Interdisciplinary MIT Study, *The Future of Nuclear Power*, Massachusetts Institute of Technology, 2003, p. 79.

[123] Gronlund, Lisbeth, David Lochbaum, and Edwin Lyman, *Nuclear Power in a Warming World*, Union of Concerned Scientists, December 2007.

[124] Travis Madsen, Tony Dutzik, and Bernadette Del Chiaro, et al., *Generating Failure: How Building Nuclear Power Plants Would Set America Back in the Race Against Global Warming*, Environment America Research and Policy Center, November 2009, http://www.environmentamerica.org/uploads/39/62/3962c378b66c4552624d09cbd8ebba02/ Generating-Failure—Environment-America—Web.pdf.

[125] Environmental Protection Agency, "Carbon Pollution Emission Guidelines for Existing Stationary Sources: Electric Utility Generating Units," proposed rule, June 18, 2014, https://www.federalregister.gov/articles/2014/06/18/2014- 13726/carbon-pollution-emission-guidelines-for-existing-stationary-sources-electric-utility-generating.

[126] Executive Office of the President, *The President's Climate Action Plan*, June 2013, http://www.whitehouse.gov/ sites/default/files/image/president27sclimateactionplan.pdf.

[127] EPA, Office of Air and Radiation, *Goal Computation Technical Support Document*, June 2014, http://www2.epa.gov/sites/production/files/2014-06/documents/20140602tsd-goal-computation.pdf.

[128] McCabe, Janet, "Understanding State Goals under the Clean Power Plan," *EPA Connect*, June 4, 2014, http://blog.epa.gov/epaconnect/2014/06/understanding-state-goals-under-the-clean-power-plan/.

[129] Nuclear Energy Institute, "Carbon Reduction Requires Big Role for Nuclear Energy," news release, June 2, 2014, http://www.nei.org/News-Media/Media-Room/News-Releases /Carbon-Reduction-Requires-Big-Role-for-Nuclear-Ene.

[130] Reeves, Dawn, "EPA May Revisit ESPS' Nuclear Provisions in Face of Industry Concerns," *Inside EPA.com*, June 13, 2014, http://insideepa.com/201406132474001/EPA-Daily-News/Daily-News/epa-may-revisit-esps-nuclearprovisions-in-face-of-industry-concerns/menu-id-95.html.

[131] Nuclear Information and Resource Service, "EPA Carbon Rule Offers Misguided Subsidies for Uneconomic, Aging and Dangerous Nuclear Reactors," news release, June 2, 2014, http://www.nirs.org/nukesandglobalwarming/ epacarbonrulepr6214.pdf.

[132] Babcock & Wilcox Company, "B&W Announces Restructuring of Small Modular Reactor Program," news release, April 14, 2014, http://www.babcock.com/news-room/Pages/BW-Announces-Restructuring-of-Small-Modular-ReactorProgram.aspx.

[133] McAuliffe, Michael, "B&W Says DOE Stops Matching Funds after SMR Program Slowdown," *Nucleonics Week*, August 14, 2014, p. 3.

[134] DOE Office of Nuclear Energy, "Small Modular Nuclear Reactors," http://www.energy.gov/ne/nuclear-reactortechnologies/small-modular-nuclear-reactors.

[135] DOE, Manifest Information Management System, http://mims.doe.gov/GeneratorData.aspx. Annual low-level waste disposal volume varies considerably, ranging during the past decade from a high of 114,841 cubic meters in 2006 to a low of 35,216 cubic meters in 2013.

[136] Blue Ribbon Commission on America's Nuclear Future, *Report to the Secretary of Energy*, January 2012, http://brc.gov/sites/default/files/documents/brc_finalreport_jan2012.pdf.

[137] DOE, *Strategy for the Management and Disposal of Used Nuclear Fuel and High-Level Radioactive Waste*, January 2013, http://energy.gov/sites/prod/files/ Strategy%20for%20the %20Management%20and%20Disposal%20of%20Used%20Nuclear%20Fuel%20and%20Hi gh %20Level%20Radioactive%20Waste.pdf.

[138] Nuclear Regulatory Commission, "In the Matter of U.S. Department of Energy (High-Level Waste Repository)," CLI-11-07, September 9, 2011, http://www.nrc.gov/reading-rm/doc-collections/commission/orders/2011/2011- 07cli.pdf.

[139] U.S. Court of Appeals for the District of Columbia Circuit, *In re: Aiken County et al.*, No. 11-1271, writ of mandamus, August 13, 2013, http://www.cadc.uscourts.gov /internet/opinions.nsf/BAE0CF34F762EBD985257BC6004DEB18/$file/11-1271-1451347.pdf.

[140] Ibid., p. 80.

[141] Hiruo, Elaine, "DOE Implements Court-Ordered Suspension of Nuclear Waste Fee," *NuclearFuel*, May 26, 2014.

[142] U.S. Court of Appeals for the District of Columbia Circuit, National Association of Regulatory Utility Commissioners v. U.S. Department of Energy, No. 11-1066, decided November 19, 2013, http://www.cadc.uscourts.gov/internet/opinions.nsf/2708C 01ECFE3109F85257C280053406E/$file/11-1066- 1466796.pdf.

[143] NRC, "NRC Approves Final Rule on Spent Fuel Storage and Ends Suspension of Final Licensing Actions for Nuclear Plants and Renewals," news release, August 26, 2014, http://pbadupws.nrc.gov/docs/ML1423/ ML14238A326.pdf.

[144] NRC, "Continued Storage of Spent Nuclear Fuel," final rule, 79 *Federal Register* 56238, September 19, 2014.

[145] The organization approved a new mission statement with the name change at its June 2010 meeting in Ghana. Seehttp://www.ifnec.org/About/History.aspx.

[146] DOE, "FY 2015 Congressional Budget Request," DOE/CF-0098, Vol. 3, March 2014, http://energy.gov/cfo/downloads/fy-2015-budget-justification; NRC, "Congressional Budget Justification: Fiscal Year 2015," NUREG-1100, Volume 30, March 2014, http://www.nrc.gov/reading-rm/doc-collections/nuregs/staff/sr1100/v30/.

[147] Senate Committee on Appropriations, "FY15 Subcommittee Reported Bill and Draft Report," July 24, 2014, http://www.appropriations.senate.gov/news/fy-2015-ew-subcommittee-reported-bill-and-draft-report.

# INDEX

## C



## D

## G



## H



## I

## Q



## R

## S



## T

## U



## V



## W

## Y



## Z